Culture-Led Urban Regeneration

The idea that culture can be employed as a driver for urban economic growth has become part of the new orthodoxy by which cities seek to enhance their competitive position. Such developments reflect not only the rise to prominence of the cultural sphere in the contemporary (urban) economy, but how the meaning of culture has been redefined to include new uses in order to meet social, economic and political objectives.

This significant book focuses on the ability of cultural investment to meet the rhetoric of social inclusion, and the extent to which it offers sustainable solutions to the problems of the city. To this end it focuses on the meanings and practice of culture-led policy within the city and its evaluation is proposed. Paddison and Miles have edited an innovative book which presents a series of diverse case studies to challenge the 'one size fits all' model of culture-led urban regeneration – a key concern being the extent to which culture-led regeneration can genuinely fulfil the expectations that policy-makers and urban commentators have of it.

This book was previously published as a special issue of *Urban Studies*.

Steven Miles is a Reader in Sociology at the University of Liverpool, UK.

Ronan Paddison is Professor of Geography at the University of Glasgow, UK and Managing Editor of the journal *Urban Studies*.

Culture-Led Urban Regeneration

Edited by
Ronan Paddison and Steven Miles

Routledge
Taylor & Francis Group

LONDON AND NEW YORK

First published 2007 by Routledge
2 Park Square, Milton Park, Abingdon, Oxon, OX14 4RN

Simultaneously published in the USA and Canada
by Routledge
270 Madison Ave, New York NY 10016

Routledge is an imprint of the Taylor & Francis Group, an informa business

Transferred to Digital Printing 2008

Typeset in Times by Techset Composition Limited

British Library Cataloguing in Publication Data
A catalogue record for this book is available from the British Library

Library of Congress Cataloguing in Publication Data
A catalogue record for this book has been requested

ISBN 10: 0-415-40038-4
ISBN 13: 978-0-415-40038-1

Contents

Urban Studies Monographs

Series Editors: Ronan Paddison, Jon Bannister, Andy Cumbers, Ken Gibb, all at the University of Glasgow

In the contemporary world cities have a renewed significance. Trends such as globalisation, neo-liberalism, new technologies, the rise of consumption and consumerism, the re-definition of modes of governance, demographic and social shifts, and the tensions to which such changes give rise, have re-defined the critical role cities play, as well as the problems arising from city life. By focusing on specific issues this series aims to explore the nature of the contemporary urban condition.

Cities of Pleasure
Sex and the Urban Socialscape
Edited by Alan Collins

Clusters in Urban and Regional Development
Edited by Andy Cumbers & Danny McKinnon

Globalisation and the Politics of Forgetting
Edited by Yong-Sook Lee & Brenda S.A. Yeoh

Employability and Local Labour Market Policy
Edited by Ronald W. McQuaid, Anne Green & Mike Danson

Introduction: The Rise and Rise of Culture-led Urban Regeneration

Steven Miles and Ronan Paddison

As a marker of its salience to the city under deepening globalisation, the most recent report in the UN-Habitat series, *The State of the World's Cities* (UNCHS, 2004), focuses on the cultural impacts of globalisation. Previous reports had singled out the problems of, and opportunities for, urban economic growth (UNCHS, 1996) and the need for improved systems of governance (UNCHS, 2001) while the 2003 study focused on slums and the upgrading of housing (UNCHS, 2003). Against this unfolding agenda, the turn to exploring the cultural dimension seems unexceptional, not least because of the self-evident linkages between globalisation, cities and cultural diversity. Further, where the report was timed to coincide with the Universal Forum of Cultures held in Barcelona in 2004, the study reflects broader concerns with multiculturalism and exclusion. While the report is concerned largely with the implications globalisation processes have on urban cultural diversity and the problems, particularly those of inequality and governance, to which this gives rise, its concern is also to show how culture has been, and can be, co-opted to address such problems of the city. Culture, it would seem, can be viewed not just as a challenge to the ability of cities to combine social justice with economic growth, but also

the source-ground around which the amelioration of such problems can be sought.

The idea that culture can be employed as a driver for urban economic growth has become part of the new orthodoxy by which cities seek to enhance their competitive position. Although, as the UNCHS report illustrates, its practice has become globalised, it is in cities in the economically advanced nations that the use of culture-driven strategies largely originated and, at least judged by the roll-call of cities that are commonly cited, that its adoption has had the most dramatic consequences both physically in transforming the urban landscape and in building their economic performance. What is remarkable here is not just the speed with which culture-driven strategies have become advocated by governments and local development agencies as a means of bolstering the urban economy, but also how their diffusion has globalised. Within the space of little more than two decades, the initiation of culture-driven urban (re)generation has come to occupy a pivotal position in the new urban entrepreneurialism. Equally, as Yeoh demonstrates (in this Review) the language of place marketing has become as integral to the Asian city as it has the European or North American city— that, more specifically, the invocation of culture has become central to the ambitions of Singapore to maintain, and enhance, its regional position as a world city.

Such developments reflect not only the rise to prominence of the cultural sphere in the contemporary (urban) economy, but how the meaning of culture has been redefined to include new uses to which it can be put to meet social, economic and political objectives. How culture has been defined, and appropriated, has not of course been static. As Bocock (1992) has argued, the definition of 'culture' has evolved, referring successively over time to the cultivation of the land, of the mind, to social development, to the meanings, values and ways of life and, most recently, to the practices which produce meanings. In more strident vein, Yudice (2003) argues that in the global era the role of culture has assumed unprecedented significance and that its redefinition as a resource has enabled it to be used as the means for resolving political as well as socioeconomic problems, including those of the city. Integral to this—and given clear expression in the city through the spread of processes of gentrification and the development of cultural forms of urban tourism—is the commodification of culture and the spread of cultural capitalism. What the relationships are between such new uses of culture and previous meanings given to the term, particularly to its part in defining established values and ways of life within the city, remain a moot point, generating contradictions around which opposition frequently gels.

The development of urban and regional policy-making is peppered with policy innovations that have progressed through a Downsian issue–attention cycle. Initial enthusiasm leading to widespread adoption, with the promise of being able to resolve an urban problem, is followed by a growing appreciation of the limitations of the paradigm and its ability to tackle complex urban and regional problems. Growth centre theory, high-rise living as a solution to the housing problems of the post-war city, New Towns, neighbourhoods—even where some such policies have lingered, or been resurrected in

new guises, the lessons of each have been to demonstrate the shortfall between initial expectations and reality. Against this history, the rise and rise of culture-led urban regeneration within contemporary policy leads to questioning whether a similar pattern may not befall it while recognising that, if it is to follow a similar cycle, we are currently at the beginnings of it. Clearly, as the next section demonstrates, the ability of the policy to deliver beyond the rhetoric, is already being questioned.

Some Key Questions

As we have argued, the rapidity with which culture has ascended the urban policy agenda has been little short of extraordinary, perhaps nowhere more so than in Britain. But the key questions remain: do we really understand the complex nature of the impact of cultural investment on our cities and how far are such decisions based on an informed analysis of how investment might change a city? More pointedly perhaps, what do such developments actually mean in terms of the lives of those people who live in that city? In short, to what extent is culture-led regeneration more about rhetoric than it is about reality?

There is no doubt that the British government has come to recognise (and arguably overestimate) the value of cultural investment to urban regeneration. This is particularly evident in Chris Smith's (2000) vision of a 'Creative Britain' in which he discusses the impact of culture-led regeneration as highlighted in the work of Comedia. In this context, Smith sees such regeneration as: an effective route for personal growth; a valuable contribution to social cohesion; of benefit to environmental renewal and health promotion; a producer of social change and

> a flexible, responsive and cost-effective element of a community development strategy [that] ... strengthens rather than dilutes Britain's cultural life, and forms a vital factor of success rather than a soft option in social policy (Smith, 2000, p. 135).

But perhaps the clearest illustration of the seriousness with which the British government has sought to promote a culture-led agenda for cities is the Core Cities initiative, set up to work in partnership with government and other key stakeholders to promote the role of the cities as the drivers of regional and national economic growth and to create internationally competitive regions. The suggestion here is that cities are the key drivers of economic change and that culture should play a key role in this process

> Culture is a source of prosperity and cosmopolitanism in the process of international urban competitiveness through hosting international events and centres of excellence, inspiring creativity and innovation, driving high growth business sectors such as creative industries, commercial leisure and tourism, and increasing profile and name recognition ... Culture is a means of spreading the benefits of prosperity to all citizens, through its capacity to engender social and human capital, improve life skills and transform the organisational capacity to handle and respond to change ... Culture is a means of defining a rich, shared identity and thus engenders pride of place and inter-communal understanding, contributing to people's sense of anchoring and confidence (Comedia, 2003).

This emphasis on culture owes much to recent debates on the relationship between culture, creativity and the city and not least the work of Richard Florida (2002) which has had a significant role in underpinning the assertion that cultural inputs translate into social and economic outputs. Florida's work resonates deeply with the regeneration agenda and although it might be argued that Florida's work is more concerned with developing an understanding of the indicative conditions favourable to the creation of urban economic growth than it is in providing a critical appreciation of them, there is no doubt that his work has had a significant impact insofar as it has captured the imagination of policy-makers. Florida argues that cities and regions should focus on promoting creativity,

and on attracting creative people, not least through their creative 'offer'. In short, for Florida, the clustering of human capital is the critical factor in regional economic growth and is the key to the successful regeneration of cities.

In Britain, such an understanding has been taken up enthusiastically at regional and national levels and not least by the Department of Media, Culture and Sport (DCMS). In the document, *Culture at the heart of regeneration*, the DCMS argues that the cultural element can become the driving-force for regeneration, as in the example of NewcastleGateshead, discussed by Miles in this Review Issue. However, there is undoubtedly a danger of exaggerating the potential impact of cultural investment. The evidential grounds for arguing, for example, that Liverpool's cultural sector will see a rapid expansion with investment of £2 billion from public and private sources and that employment in the cultural sector will grow by at least 14 000 jobs as a result of the award of the Capital of Culture 2008 title, remains at best limited (DCMS, 2004). Indeed, Jones and Wilks-Heeg argue that the model of regeneration promoted by this sort of approach is inherently misleading to the extent that

> Current trends suggest precisely the scenario of a rapidly regenerating and gentrifying urban core surrounded by a ring of intensely disadvantaged residential areas (Jones and Wilks-Heeg, 2004, p. 357).

Despite acknowledging that they have a limited evidential basis for proving the benefits of culture-led regeneration, the DCMS continue to make claims as to its potential. This is an issue that will be raised throughout the Review Issue. How do we go about understanding the impact of culture-led regeneration in a way that provides a more balanced understanding of its pros and cons? The nature and role of cultural policy have in recent years been transformed—notably through the ideological delegitimisation of state intervention and public-sector arts and media

In the field of culture and cultural policy, civil society and the public sphere of rational-critical debate represent the possibilities of challenge and resistance to corporations that are only accountable to their shareholders and governments that submit too readily to corporate interest (McGuigan, 2004, p. 60).

Indeed, it is worth reflecting on the role of the DCMS in promoting the cultural case in the context of citizenship and, more specifically, 'cultural citizenship'. The impact of culture-led regeneration is clearly closely tied up to a localised sense of place. Government discourse around culture certainly acknowledges this fact, but it remains doubtful as to whether local issues are given full rein when broader economic ones appear to be so much more immediate. This reflects the concern recently expressed by Culture North West (2004) on their website that the primary focus on new landmark investment as the route to regeneration should not be supported as being the key driver for culture-led regeneration: "Our work suggests that a finer grain, more subtle and locally finessed approach is more appropriate for much of the region". There are undoubtedly some signs that the government has been increasingly willing to move in this direction in recent years. The influential Urban Task Force, chaired by the architect Lord Rogers, said in its 1999 report that British cities were 'way behind' those in Holland, Germany and Scandinavia in terms of the quality of urban life and the built environment and that in turn improvements in design were vital for an 'urban renaissance' to reverse the abandonment of inner cities and to protect the countryside from sprawling development (Urban Task Force, 1999).

The role of culture in the above process remains uncertain. Ward (2002) goes as far as to describe the impact of an 'enduring myth', the myth that culture has to be a good thing and that there may be money in it. Wilks-Heeg and North (2004) point out that local economic development strategies have increasingly identified cultural and crea-

tive industries as a key growth sector in urban and regional economies. They go on to point out that the Tate Modern is estimated to be worth £100 million whilst supporting 3000 jobs in London. But there is a danger here, as Stevenson (2004) points out, that cultural planning has come more and more to be concerned with "intervening and achieving outcomes that relate to a conception of culture as a civilising process that is not dynamic, flexible and situational, but linear and linked to a set of clearly defined political and governmental objectives" (Stevenson, 2004, p. 125). Stevenson goes on to discuss the way in which cultural planning meshed with the Third Way objective of seeking to transcend the welfare consensus of the old Left in favour of a social democratic schema; characterised above all by the focus on social inclusion rather than social justice

> In the language of the Third Way, the 'social' of social inclusion has become synonymous with the economy to such an extent that participation in society (full citizenship) can only be achieved through participation in the economy ... There is no scope in the rhetoric of the Third Way to assess or address the causes of social exclusion or disadvantage. . . . To be more specific, there is no language for discussing the extent to which the ability of the top to 'splinter off' from society actually depends on the structural 'exclusion' of the bottom (Stevenson, 2004, p. 126).

From this point of view, the degree to which culture itself is implicated in the reproduction of inequality is largely neglected as a result of the apparent fusion of the social, the economic and the cultural. Social inclusion therefore becomes determined by an individual or social group's relationship to the marketplace and, by implication, their role as consumers. The problem with this is that all too often there appears to be an assumption that the rehabilitation of the urban will automatically revitalise the public sphere. Stevenson's argument here is that cultural planning is premised on a kind of strategic pun that sees cultural activity and the creative industries as the scaffolding upon which vibrant urban economies can be established, whilst these very same strategies and outcomes are touted as a means of developing the cultural capital of the local population in a way that addresses social exclusion

> These competing objectives have collapsed in on themselves as undifferentiated elements of holistic cultural planning. They are explained and legitimated in terms of each other in ways that are not only tautological, but also disguise significant political motives and assumptions (Stevenson, 2004, p. 128).

The danger then, as Stevenson expresses it, is that if cultural planning is to be a success, culture needs to mean something, but it can and should not be expected to mean everything.

The rhetorical promotion of culture as a sort of an economic panacea is profoundly shortsighted and indeed underestimates the value of culture for the people of a locality. Several of the articles presented here touch on the potential of cultural investment to "refresh the local soul as well as the local economy" (Ward, 2002, p. 7). The single most dangerous aspect of cultural investment is that it simply does not sit comfortably in the context for which it is intended. This tendency is discussed by Jayne (2004) who looks at misfiring attempts to use the arts and cultural reproduction for urban regeneration in Stoke-on-Trent. Critiquing the suggestion that cultural investment, notably in the creative industries, can attract post-industrial jobs and encourage people back to living in city centres, whilst generally improving the urban quality of life, Jayne argues that it is important that critical rigour is applied to the ways in which creative industries development has become aligned with regeneration in our cities

> Unlike many other Western cities, Stoke-on-Trent remains overly dominated by working-class production and consumption cultures. The city is thus, in a sense, rendered illegible to post-industrial

businesses, tourists and to the many young people who leave the city in search of the more dynamic economic and cultural opportunities offered in other cities (Jayne, 2004, p. 208).

The key focus here should therefore not be on whether cultural investment works, but on the degree to which it works for diverse social groups. This is an issue raised especially effectively in this Review Issue by Evans who considers the data-collecting implications of cultural investment. The emphasis on economic impacts produces headline-grabbing data about the raw potential of cultural investment, but it says next to nothing about the long-term sustainability of culture-led regeneration. The promotion of cultural investment in this way may actually prove to be self-defeating, with the realisation that such data cannot actually prove anything about the causal impact of cultural investment. As things stand, such 'research' acts as little more than a form of promotional activity. As Johnson and Thomas put it

> There is . . . a need to pursue an altogether wider agenda on the economic impact of the arts. This agenda would focus more on the measurement and valuation of the impact of the activities of the arts sector on the enjoyment, appreciation and human capital of participants, and on those whom they influence—in other words, the cultural impact . . . Such research would . . . be much closer to the underlying rationale for the arts and would provide an altogether more satisfactory basis for policy evaluation, as it would seek to capture those benefits that are not reflected in market transactions (Johnson and Thomas, 2001, p. 216).

It is also essential that such a move is premised on sound methodological grounds, rather than depending upon the more journalistic approach that tends to be associated with research on the sector undertaken by think-tanks which can usually and ironically depend upon more direct links with policy-makers (Minton, 2003).

Questions focusing on the ability of cultural investment to meet the rhetoric of social inclusion are accompanied by the extent to which it offers sustainable solutions to the problems of the city. This value is expressed in economic terms—the extent to which culture-led regeneration is able to underwrite the economic growth of the city. Given the variety of different types of investment and the play of contextual factors, the ability to generalise on the longer-term economic impacts of cultural regeneration is constrained. No doubt the extent to which such strategies offer the prospect of durable solutions to the needs of the city differs between the type of city as well as it does between the various types of cultural investment. Some types of investment, the promotion of the city for cultural tourism, may be vulnerable to exogenous factors that local development agencies are hardly in a position to influence. Other types, such as the holding of mega-events are by their nature transitory, but are based on a rationale which has imputed longer-term effects. Still others— the development of the city as centres of retail consumption able to attract custom on a national or international basis—may have a limited life-span where emulation by rival cities alters the map of retail attractiveness, or where cities become victims of fashion. Among those cities in which culture-led regeneration has had the more dramatic consequences on their external (and self-) image, what can be termed a 'feelgood' factor becomes part of the justification of the strategy. But by itself such a justification is hardly likely to have durable effects. This is not to deny that, where culture-led regeneration has become intertwined with the construction of more positive images of the city, it does not raise its attractiveness as a place in which to invest and live, along the lines argued by Florida (2002) and others. But being able to show it has such intended consequences is another question.

Further, sustainability needs to be considered for its social and political implications, besides that of the economy. Here, sustainability refers to the ability in which

cultural policy is able to be inclusive and the extent to which this contributes to defining the political legitimacy of collective decision-making. As the use of the arts in disadvantaged neighbourhoods demonstrates, the sensitive use of cultural policy can contribute positively to the ways in which residents of otherwise stigmatised places identify with them. Nor is it assumed that such social impacts are not unconnected to the economic—that cultural policy, in other words, can contribute to social cohesion, itself assumed to be important for the competitiveness of the city. In fact, such connections are at the least debatable (Boddy and Parkinson, 2004), emphasising the caution necessary in extrapolating the ability of policies aimed at cultural and social change as also leading to desired economic outcomes. Meanwhile, as long as funders and policy-makers continue to obsess about raw indicators, they will also continue to underestimate the degree to which cultural developments succeed depending on how well they engage with local communities and cultures. As things stand, there is a perennial danger that what we are constructing here is a "manufactured culture, drawn up by regeneration specialists and regional redevelopment advisers, at its most blatant" (Hunt, 2004, p. 350). Nonetheless, it remains the case that the careful analysis of the impact of culture-led regeneration offers hope as to the potential benefits to be had from strategically balancing economic and cultural imperatives effectively in the name of urban regeneration.

The Rationale of the Book

Summarising the previous argument, two key, interrelated questions arise from the current deployment of culture-led urban regeneration: the critical appreciation of what it aspires to achieve; and, how to evaluate whether it is meeting these expectations. The papers in this volume organised around these two themes. The collection does not attempt to provide a comprehensive overview of how culture intersects with the urban. Thus, certain issues normally included within statements on the urban cultural economy—the development of cultural industries, the museumisation of the city, or the development of urban cultural tourism, for example—have not been singled out for separate consideration. Rather, our intentions are more focused on the meanings and practice of culture-led policy within the city and its evaluation. In a rapidly developing policy field and against a burgeoning academic response to it, the contributions are hardly likely to be the last word on the subject, but offer considered reflection on the current state of play.

References

AUGÉ, M. (1995) *Non-places: Introduction to the Anthropolgy of Non-places*. London: Verso.

BASSETT, K., GRIFFITHS, R. and SMITH, I. (2002) Testing governance: partnerships, planning and conflict in waterfront regeneration, *Urban Studies*, 39(10), pp. 1757–1775.

BOCOCK, R. (1992) The cultural transformations of modern society, in: S. HALL and B. GIEBEN (Eds) *Formations of Modernity*, pp. 229–274. Milton Keynes: The Open University and Polity Press.

BODDY, M. and PARKINSON, M. (Eds) (2004) *City Matters*. Bristol: The Policy Press.

COMEDIA (2003) Releasing the cultural potential of our core cities: culture and the core cities (http://www.corecities.com/coreDEV/co media/ com_cult.html).

CULTURE NORTH WEST (2004) (http://www. englandsnorthwest-culture.com/cultural/).

DEPARTMENT OF CULTURE, MEDIA AND SPORT (2004) *Culture at the Heart of Regeneration*. London: DCMS.

FLORIDA, R. (2002) *The Rise of the Creative Class*. New York: Basic Books.

HUNT, T. (2004) *Building Jerusalem: The Rise and Fall of the Victorian City*. London: Weidenfeld and Nicolson.

JAYNE, M. (2004) Culture that works? Creative industries development in a working-class city, *Capital & Class*, 84, pp. 199–210.

JOHNSON, P. and THOMAS, B. (2001) Assessing the economic impact of the Arts, in: S. SELWOOD (Ed.) *The UK Cultural Sector*, pp. 202–216. London: Policy Studies Institute.

JONES, P. and WILKS-HEEG, S. (2004) Capitalising culture: Liverpool 2008, *Local Economy*, 19(4), pp. 341–360.

McGUIGAN, J. (2004) *Rethinking Cultural Policy*. Maidenhead: Open University Press.

MINTON, A. (2003) *Northern Soul: Culture, Creativity and the Quality of Place in Newcastle and Gateshead*. London: DEMOS.

SASSEN, S. (1991) *The Global City: New York, London and Tokyo*. Princeton, NJ: Princeton University Press.

SMITH, C. (2000) *Creative Britain*. London: Faber and Faber.

STEVENSON, D. (2004) 'Civic gold' rush: cultural planning and the politics of the Third Way, *International Journal of Cultural Policy*, 10, pp. 119–131.

UNCHS (HABITAT) (1996) *An Urbanizing World: Global Report on Human Settlements* 1996. Oxford: Oxford University Press/UNCHS (Habitat).

UNCHS (HABITAT) (2001) *Cities in a Globalizing World: Global Report on Human Settlements* 2001, London: Earthscan/UNCHS (Habitat).

UNCHS (HABITAT) (2003) *The Challenge of Slums: Global Report on Human Settlements*. London: Earthscan/UNCHS (Habitat).

UNCHS (HABITAT) (2004) *The State of the World's Cities 2004/2005: Globalization and Urban Culture*. Nairobi: UNCHS; and London: Earthscan.

URBAN TASK FORCE (1999) *Towards an Urban Renaissance: Report of the Urban Task Force*. London: Office of the Deputy Prime Minister.

WARD, D. (2002) The Guggenheim effect, cities reborn: the challenge of an urban renaissance, *The Guardian*, 30 October.

WILKS-HEEG, S. and NORTH, P. (2004) Cultural policy and urban regeneration: a Special Edition of Local Economy, *Local Economy*, 19(3), pp. 305–311.

YUDICE, G. (2003) *The Expediency of Culture: Uses of Culture in the Global Era*. Durham, NC: Duke University Press.

Deconstructing the City of Culture: The Long-term Cultural Legacies of Glasgow 1990

Beatriz García

Introduction

The phrase 'culture-led urban regeneration' has grown from an interesting alternative to urban development policy into a core strategy in an increasing number of cities and regions world-wide. From the US-based 'festival marketplace', "a formula for redeveloping derelict waterfront sites which pivots on consumption, entertainment and spectacle" (Stevenson, 2003, p. 141) to the increasingly adopted 'cultural planning' approach, aiming at "nurtur[ing] and promot[ing] local cultural activity in the city" (p. 141), culture-led regeneration is a core priority in urban centres as diverse as Barcelona, Montreal and Singapore.

This paper discusses a particular instance of such developments in Europe: the European City/Capital of Culture programme (ECOC).

The ECOC started as a rather sanguine EU initiative but has been transformed into what is perceived as an attractive catalyst for cultural regeneration, generating enormous expectations in cities from countries as diverse as the UK, the Netherlands and Greece. The programme is an interesting case study because it has evolved over the past couple of decades in parallel with the growing debate around definitions and uses of culture-led regeneration and has touched all EU countries in turn. It is a programme that did not originate from clearly structured guidelines as to what would constitute a 'European City/Capital of Culture'. Indeed, its history has been one of adapting to the needs and demands of those cities hosting it rather than imposing a prefigured model of urban cultural policy. In this context, the issue at stake is whether this programme has managed to address successfully the expected outcomes of culture-led regeneration.

A recent report by Evans and Shaw (2004) reviews the current state of evidence on culture's contribution to regeneration in the UK in which, they note, culture-led regeneration is one of three models.[1] The report establishes that there is a wealth of approaches to 'impact' measurement (the most common term used to study the contribution of cultural activity to other objectives), with tests particularly developed in the areas of environmental (or physical) and economic impact assessment (Evans and Shaw, 2004, p. 6).[2] However, they note important weaknesses due to the lack of evidence about long-term legacies and the limited understanding of social and, particularly, cultural impacts as opposed to economic and physical impacts (pp. 31–32, 57–59). Contrasting with the poor state of longitudinal evaluation techniques, the notion of long-term impact and 'sustainability'—understood as beneficial inputs for the city and its inhabitants that are able to survive and develop beyond five years— is increasingly seen as a key measure of success within urban regeneration programmes (Bianchini, 1999; Egan, 2004; Frey, 1999; Urban Task Force, 1999). This is because, as argued by Evans and Shaw, "regeneration is a fragmented process that takes place over several years, perhaps a generation or more" (p. 57) and "the complexity of the process of regeneration makes it hard to attribute an effect to a cause, *particularly in the short term*" (p. 29; original emphasis).

Cultural impacts are also increasingly valued as a desired effect in their own terms

[Beyond physical, economic and social impacts,] researchers ... have begun to look ... at a fourth type of impact—*cultural impact*. This term is already being used to describe two rather different effects. One is the impact on the cultural life of a place. For example, the opening of a gallery where there was none before ... The other use refers to the impact of cultural

activity on the culture of a place or community, meaning its codes of conduct, its identity, its heritage and what is termed 'cultural governance' (i.e. citizenship, participation, representation, diversity) (Evans and Shaw, 2004, p. 6; original emphasis).

It is thus through assessing the long-term cultural impacts—or sustainable cultural legacies—of the ECOC that this paper aims to evaluate the success of the programme as a model of culture-led regeneration. The paper focuses on the experience of Glasgow in 1990, the first city to be widely acclaimed for showing how the designation might be appropriated to underpin the wider project of regeneration. The analysis of this experience 15 years on will provide a basis to argue the core benefits of culture-led regeneration and will help to assess whether the experience led to cultural legacies that were sustainable in the long term.

The paper reflects on the findings of a three-year project conducted by the Centre for Cultural Policy Research (CCPR) at the University of Glasgow. The project takes a similar stance to that reported by Bailey *et al.* (2004, p. 47) when arguing the importance of longitudinal studies in providing "an in-depth understanding of geographical and historical specificities" which is in turn the only way to "understand the way in which cultural regeneration potentially strengthens existing sources of identity rather than imposing new ones". In this paper, the effect of regeneration on local identities, including citizens' self-perception and the perception of the place they live in, is seen as a key cultural impact and one with the potential to be sustained in the long term.

The CCPR project is based on retrospective methods of study, tracing the progression of media and personal discourses on the city's approach to regeneration, event hosting and urban cultural policy in the two decades that separate Glasgow's 1986 ECOC nomination from the 2003 nomination of Liverpool as the next UK Capital of Culture in 2008. The paper aims to show the value of assessing soft indicators such as media and personal discourses as an approach to measuring cultural impacts and legacies and a means of complementing the analysis of other more visible and commonly assessed impacts such as the event's effect on the city's infrastructure, levels of employment and visitor attraction (see Myerscough, 1991, 1994). The paper argues that the most sustainable legacies of Glasgow 1990 have been cultural rather than physical or economic but that they have not been properly assessed over time and are often dismissed as purely anecdotal, partly due to their subjective nature. This situation leads to the conclusion that it is necessary to keep developing longitudinal and qualitative impact and legacy evaluation techniques. In Glasgow's case, understanding the key to its successful experience as an ECOC and the reason why it is still considered a key referent 15 years on will not result from a purely economic and environmental analysis but rather from investigating the formal and informal narratives created around such an event. This paper aims to demonstrate that these are the most important sources of current pride and belief in the city's potential as a creative centre and are thus its more sustainable legacy.

The City/Capital of Culture Programme and Culture-led Regeneration

The ECOC programme was conceived in 1983 by Melina Mercouri, then Greek Minister for Culture. The purpose of the programme was to give a cultural dimension to the work of the European Community (now the European Union) at a time when it did not have a defined remit for cultural action and to celebrate European culture as a means of drawing the community closer. As argued by Evans (2003, p. 425), the ECOC is an example of the European Union's progressive shift from an almost-exclusive focus on the creation of common market (free trade) instruments and regional development, into more localised city-based initiatives.

Evans suggests that a defining characteristic of the ECOC is that it

has acted as an effective 'Trojan horse' by which structural economic adjustment policies and funding have been diverted into

arts-led regeneration ... generally bypassing national and even city cultural and economic development policy. ... The use of culture as a conduit for the branding of the 'European Project' has added fuel to culture city competition, whilst at the same time celebrating an official version of the European urban renaissance (Evans, 2003, p. 426).

The ability of such a programme to surpass local cultural policies has been contested by some (Myerscough, 1994, p. 24). However, there is little question of the programme's effect on increasing city competitiveness and promoting culture-led regeneration agendas in an expanding Europe and within the UK in particular (see Cogliandro, 2001; Davies and Russell, 2001; Gulliver, 2002; Palmer/Rae Associates, 2004).

The selection of Glasgow in 1990 marked the start of the ECOC as a catalyst for urban regeneration. Initially used as an opportunity to reinforce the status of prestigious European cultural centres—such as Athens (1985), Florence (1986), Amsterdam (1987), West Berlin (1988) and Paris (1989)—after Glasgow, the title has been integrated within medium-to-large regeneration projects and used to promote emerging cultural assets in capital, second and third cities alike (see García, 2004a, p. 319). Copenhagen, Thessalonica, Stockholm, Weimar, Porto, Graz, Genoa and Lille are some examples of cities that have linked the ECOC to ambitious urban and regional regeneration strategies, with recent studies showing these cities' dedication to above-average levels of funding to operate specially designed programmes of activity for up to one year (€50–73 million as opposed to €40 million on average by ECOC hosts between 1995 and 2004) and/or fund capital projects (€150–230 million as opposed to €105 million in the same period) (Palmer/Rae Associates, 2004, pp. 85–89).

However, overall, the ECOC programme reveals a series of weaknesses that mirror many of the still unsolved tensions in European approaches to culture-led regeneration. One important problem is the extremely low level of European funding allocated to the programme. Although the EU has increased its budget from an initial average of €120 000 per city to the current allocation of €500 000 since 2001, these amounts are clearly insufficient to support a full programme of activities, especially in the face of growing public expectations and ever-tougher interurban competition. A recent assessment of ECOC sources of funding from 1995 to 2004 reveals that EU support accounts for barely 1.53 per cent of total income while national governments cover up to 56.84 per cent of all costs and city and regional authorities up to 31 per cent (Palmer/Rae Associates, 2004). A further problem has been the lack of any systematic monitoring of the ECOC programme as it has unfolded, limiting the extent to which cities have been able to learn from one another. As noted by García

> Despite attempts at creating platforms to share know-how ... there is no formal monitoring mechanism in place. As such, the information available about ECOC experiences relies entirely on the willingness of host cities to produce final reports. ... Comprehensive reports are ... scarce and mostly restricted to the assessment of immediate impacts, without a follow-up study in the medium to long term. The resulting effect is the creation of virtually unquestioned 'myths' about the value of hosting the title that cover up the lack of serious attempts to learn lessons from the experience and establish replicable models of successful and ... sustainable culture-led regeneration. (García, 2004a, p. 321).

A detailed study of one of the most celebrated ECOC experiences will help to uncover the limitations as well as the successes of the programme in creating long-term cultural legacies.

The Experience from Glasgow 1990 to Liverpool 2008

To understand the influence of the ECOC on approaches to culture-led regeneration, the experience of the UK is particularly revealing. Beyond the claim that Glasgow was the first city clearly to associate the ECOC with urban regeneration, the UK government, 'New Labour' in particular, is supportive of the use of culture as a tool for regeneration, with the Department of Culture, Media and Sport (DCMS) framing it as one of its key objectives and embarking on a major consultation programme, 'Culture at the Heart of Regeneration' (DCMS, 2004).

Furthermore, in 2003, Liverpool was nominated to be the second UK city to hold the title. The level of debate surrounding this decision was well above what is common in other European countries, where the title is often allocated to one city without an open competition process (for example, in Spain, Greece and France) and the level of public debate and media interest tends to be far less.[3] The analysis of UK press coverage and key promotional messages launched by competing candidates in the bidding period (2002–2003) indicates that Glasgow 1990 was used as a key reference-point and presented as the role model to be replicated in 2008. This situation prompts the question of whether Glasgow is indeed a role model for culture-led regeneration in the first place, demanding a careful consideration of what this means by identifying the elements that are more commonly referred to in existing claims of success and by being aware of the origin of these claims and their context.

The Glasgow Model: A 'Success' Story?

Glasgow is a good case to study the evolution, successes and failures of the ECOC programme because it represents a turning-point in the initiative. It was the first city to win the title after an open national competition, the first to have more than three years to plan the event, the first to gather substantial public and private support to fund event-specific initiatives and the first to understand the potential of the ECOC as a catalyst for urban regeneration through culture. However, Glasgow is also a case framed by controversy, with strong claims made against the use and abuse of cultural celebrations in a city that has yet to resolve a remarkable divide between the wealthy and the poor (Boyle and Hughes, 1991, 1994; Mooney, 2004). The city remains socially divided both because of the legacy of its industrial past and the nature of the city's gentrification (see MacLeod, 2002; Mooney and Danson, 1997). There exists a wide range of materials produced either in support of or against Glasgow's approach to culture-led regeneration. However, it is rare to find accounts capable of integrating both perspectives and, as noted below, a majority of claims are based on work produced in the early 1990s, thus overlooking any possible empirical update after 1995.

The claim that Glasgow 1990 is a success story and the model to follow 15 years on is strongly reflected in the national press coverage surrounding the bidding process to host the ECOC'2008 in the UK. A review of 350 articles debating the bid between January 2002 and June 2003 (point of final nomination) shows 90 per cent of positive references to Glasgow 1990 with an emphasis on references to the city's image transformation (31 per cent), followed by references to the event's positive economic legacies in general (19 per cent) and the growing levels of tourist visits in particular (17 per cent). These extremely high levels of positive coverage must indeed be understood as an effect of the biases that tend to frame major event bidding processes at a national level and, in particular, within the local press of respective candidate cities. As shown in the section on 'Image legacies', media references to Glasgow 1990 have been increasingly positive in the post-event period, but they are complemented by significant levels of negative and neutral coverage.[4]

Further to the press hype surrounding the 2008 UK nomination, references to Glasgow as an ECOC role model also emerge from a wide range of specialist policy and planning publications. These range from immediate celebratory accounts, which tend to emphasise Glasgow's remarkable transformation since the early 1980s thanks to the development of strong public–private partnerships and city

marketing strategies (Rae, 1993; Sayer, 1992; Wishart, 1991); to medium-term appraisals establishing Glasgow's pioneering role in culture-led regeneration (Myerscough, 1994; Kirkpatrick, 1996); to long-term recollections attempting to identify the event's key legacies (Davies and Russell, 2001; Gulliver, 2002; MacDonald, 2002; Wood, 2002). Glasgow's reputation is also supported by the widely cited *Monitoring Glasgow 1990*, a report produced by John Myerscough (1991) with funding from the city and regional authorities. This is one of the most comprehensive studies of an ECOC hosting process to date and another reason for Glasgow's predominance in debates about the regenerative potential for the city of securing the ECOC. In most if not all of these publications, the praise concentrates on the city's image transformation from grim industrial centre to attractive creative hub, including the growth in leisure and business tourism that resulted partly from this image transformation. This suggests a predominance of the economic rationale to justify Glasgow's success and a trend towards overlooking its wider social and cultural implications.

In contrast, references to Glasgow within academic publications are often dismissive of the rhetoric surrounding 1990 (Booth, 1996; Boyle, 1997; Boyle and Hughes, 1991, 1994; Mooney and Danson, 1997; Spring, 1990). It is also common to find extremely critical accounts within leftist magazines and related publications (Kemp, 1990; McLay, 1988, 1990). From the mid 1990s onwards, these accounts have spread within activist sites in the World Wide Web (Clark, 2002; Richards, 2002). The main criticism emerging from such publications is Glasgow's failure to resolve its social problems. Typically, 1990 success was seen in economic terms generating low–wage jobs and benefiting élites—political, corporate, cultural or otherwise—while doing little to alleviate the city's underlying structural economic problems. In this frame and in line with David Harvey's remarks (1989),[5] Glasgow's ability to put on a major event and gather international acclaim is considered a mask

aiming to hide the enduring, embedded problems and contradictions resulting from
decades of poverty and related housing,
health and nutrition problems. These contradictions support Boyle and Hughes' arguments (1991, 1994) about the highly
politicised nature of the 1990 event.

Interestingly, there have been practically no
published academic studies on the Glasgow
1990 experience beyond the mid 1990s.
Liverpool's nomination has already triggered
reactions from academics, some of which
stem from a Marxist tradition that openly
rejects the rhetoric of urban renewal and/or
regeneration through major events (see
Mooney, 2004). This has resulted in an irre-
concilable opposition of arguments. In this
context, this paper follows from earlier publications resulting from the CCPR project
(García, 2004a, 2004b) that aim to act as a
bridge between the more celebratory and the
more confrontational accounts by engaging
with current arguments about the benefits of
event-led regeneration, at the same time as
critically assessing its cultural impact on
local communities. The issue at stake here is
whether we accept claims of success that are
exclusively grounded in economic terms or
whether we also take into account a cultural
discourse that appreciates cultural legacies
beyond economic returns and symbolic
effects beyond purely social benefits (see
Bailey et al., 2004; Evans and Foord, 2003;
Mommaas, 2004).

*Methodological Challenges: Assessing the
'Cultural' Legacies of Culture-led
Regeneration*

In formulating the CCPR project, an important decision was to move beyond a focus on
economic outputs and assess instead the cultural dimension of urban regeneration, understood as the effect of regeneration on the
culture of a place or community (see Evans
and Shaw, 2004, p. 6, cited in the introduction). Yet, a major challenge has been
finding appropriate methodologies able to
assess such cultural effects in a field otherwise

dominated by economic and environmental
impact studies.

The general underpinnings of the preference for a 'cultural economic policy' (Kong,
2000) have been widely documented
(Greenhalgh, 1998; Griffiths et al., 1999;
Peacock and Rizzo, 1994) but the phenomenon is most evident in the context of
regeneration discourses (Bianchini, 1990;
Evans and Foord, 2003; Mommaas, 2004).
The recent report to the DCMS by Evans
and Shaw (2004) notes that, although
culture-led regeneration strategies are now
commonplace, methods to assess the cultural
impact of such regeneration are severely
underdeveloped. This is because, in addressing the concerns of their funders (typically
government agencies and/or corporations),
most assessment studies focus on measuring
economic, physical and—on occasion—
social impacts. However, little explanation is
given about the *cultural* benefits of such a
process. This suggests that culture is being
used as an instrument for other ends and
that, given the growing interest in its urban
application, it is proving successful at achieving them—but the understanding of cultural
impacts and legacies remains unaddressed.

In Glasgow's case, a common claim is that
the most successful and sustainable legacy of
hosting the ECOC was precisely of a cultural
and symbolic nature: the transformation of
Glasgow's image. However, in line with
Evans and Shaw's claims, the value of such
a symbolic legacy is justified in more tangible
economic terms: the growth in the number of
tourists, business relocation and inward
investment. These economic justifications
fail to explain the cultural and social dimensions of image transformation. This was the
point of departure for the CCPR research
project.

Indeed, in trying to evaluate the worth of a
'symbolic legacy', the research had to face the
challenge of measuring supposedly intangible
elements, which is seen as a common problem
of cultural impact studies (Evans and Shaw,
2004, pp. 28–29). To address this difficulty,
image change was measured by assessing
the progression of personal, institutional and

media produced narratives about Glasgow as a city to visit, work in or live in, in the context of the ECOC'1990 or its aftermath. The measurement involved the identification of key indicators such as references and attitudes towards the city's promotional or marketing strategy and related discourses on city development; published opinion or personal perception about its quality of life and/or liveability; perceptions on the levels and appropriateness of cultural provision, facilities and infrastructure; opinions about the effect of 1990 on local confidence, creative development and creative outputs; and community participation in the arts and other cultural endeavours, civic involvement and/or cultural citizenship.[6]

These narratives were initially traced by undertaking an extensive content analysis of media coverage on Glasgow 1990 published between 1986 and 2003. The analysis of media narratives was then contrasted with the narrative emerging from official promotional discourses or 'city marketing' strategies. This second area of study is particularly relevant as, beyond pioneering an era of culture-led regeneration, Glasgow has been cited as a role model for exploiting city marketing as a core component of urban regeneration (Paddison, 1993). The study of city marketing strategies involved the review of archives and documentation produced by local authorities and the Glasgow Tourist Board between 1983 and late 2003. Finally, media and marketing discourses were placed in the context of the personal narratives of special interest groups representing Glasgow's cultural, political and business worlds. The names of individuals or institutions relevant to the project were identified through the documentation review, media content analysis and initial interviews with key informants. The criteria for selection were a direct involvement (as politicians, managers, artists, opinion leaders, etc.) with the ECOC and its immediate aftermath and/or a current role within institutions, venues or activities considered a direct or indirect legacy of 1990. The assessment of personal accounts was undertaken through one-to-one

interviews with 45 people and 7 focus groups with an average of 5 people each.[7]

These three research strands are strongly interrelated and were often developed simultaneously, using the findings emerging from one area to inform the assessment of the rest. However, due to space limitations, only the first and third strands will be explored in some detail here. The analysis of press coverage is presented below as 'image legacies', reflecting on the claim that media narratives have been one of the main drivers of Glasgow's image transformation. The analysis of personal accounts or perceptions of Glasgow is understood as a form of 'identity legacies' and presented in the following section to explain the impact of Glasgow 1990 on the lives of individuals and/or the local interests they represent.

Image Legacies: Tracing Media Narratives about Glasgow's Image and Quality of Life

In attempting to undertake a longitudinal study about a past event, the research has had to overcome the limitations inherent in any retrospective study. These are the difficulty in comparing current data (over 14 years later) collected according to the project's priorities, with data collected before, during and shortly after the event year by other researchers guided by rationales which do not necessarily respond to our current aims. In this sense, the extensive monitoring report produced by John Myerscough (1991), despite providing a thorough quantitative assessment of materials contemporary to the event, is not particularly useful for the project. This is because its horizons are limited to assessing impacts during and immediately after 1990, and the approach emphasises the event's economic impacts rather than social or cultural effects. To address this limitation, the collection of current personal and institutional narratives has been complemented by the review of media coverage contemporary to the extensive period covered by this project.

The press articles collected offer not only an insight into the narratives produced at different points in time, but also an opportunity to compare how the publicly stated views from those directly involved (as politicians, event managers, artists, anti-ECOC campaigners, etc.) have changed in the process. Indeed, many of the people we have interviewed in 2002 and 2003 had also been interviewed in the media or written their own articles between 1986 and 2001. This provides a basis to assess the evolution of their perceptions of the city and/or their own aspirations and helps to identify the possible contextual factors influencing such an evolution.

Our interest in media discourses is also a result of our aim to assess Glasgow's image change as a key factor in the formation of local identities and thus as a reflection of the event's cultural impacts. Media discourses are not seen as a direct reflection of the reality of a place but rather as an influencing factor on public opinion and thus a relevant source of information for any local debate about the state of the city and its future. As argued by Jesús Martín-Barbero

> Communication media play [a role] in cultural change and the anthropological span of the change produced through communication . . .. As suggested in the title of my book, 'From media to mediations', I try to think not only about [the media as] means but also [as] ends: [that is, about] how they are changing the constitution and recognition of collective identities, and the incidence of the media as well as other processes of communication on the reconstitution of such identities (Martín-Barbero, 1991, p. 4; trans. from Spanish).[8]

The analysis of media narratives was conducted through a content analysis of 5700 articles published mainly in UK newspapers between 1986 and 2003. Two phases were identified in this study. The first, from 1986 to 1991, was denominated the '1990 period', as the coverage was dedicated to discuss 1990 issues as they unfolded, both in the lead-up or indeed during the event itself. This period was also distinguished by the articles' format, as they were only available

in print and kept as an archive of 1990 at the Mitchell Library in Glasgow. These articles (from now on 'clippings') had been originally gathered by the Glasgow 1990 press office and were assessed by the research team to ensure their suitability for the research and the validity of the selected sample. The second phase, from 1992 to 2003, was denominated the 'post-1990 period' and was characterised by the retrospective approach taken by journalists towards 1990 issues, which resulted in a progressive emphasis on identifying (or noting the lack of) event impacts and legacies. Clippings from this period were available in electronic format and were selected by undertaking key-word searches in the database 'Lexis-Nexis'.

Following established methods of content analysis (see Berelson, 1952; Bryman, 2001), the clippings were individually coded according to a series of categories reflecting the key indicators we were using to measure image change. These included profile categories such as date of publication; newspaper title; geographical remit (UK national, Scottish national, local paper, international); newspaper type (broadsheet, tabloid); article length and article format (editorial, news, feature, etc.); and qualitative categories, such as thematic focus and paper attitudes towards respective themes. The complexity of such a process of categorisation, in particular, the thematic and attitudinal assessment, is fully discussed in a working paper by Reason and García (2003).[9] Here, the discussion will focus on a selection of relevant findings resulting from the indicators that we feel best address the paper's objectives. These are the geographical provenance of papers, the main themes covered and respective attitudes towards them.

An assessment of the geographical provenance of newspapers helps to establish the extent to which certain themes are of local, national or international interest. This is also useful to contextualise the origin of current claims about Glasgow's success or failure and to assess whether these claims have predominantly a local basis, framed by household papers and opinion-leaders, or whether they have been mostly influenced by the views of outsiders.

The range of papers and magazines covering Glasgow 1990 was extremely large. For ease of analysis, papers were grouped in categories based on their provenance (see Figure 1). The coverage in the 1990 period is clearly dominated by the two main Glasgow-based papers, *The Herald* and the *Evening Times* (42.6 per cent). Combined with the Edinburgh-based *Scotsman* and local and regional tabloids, the Scottish press dominated

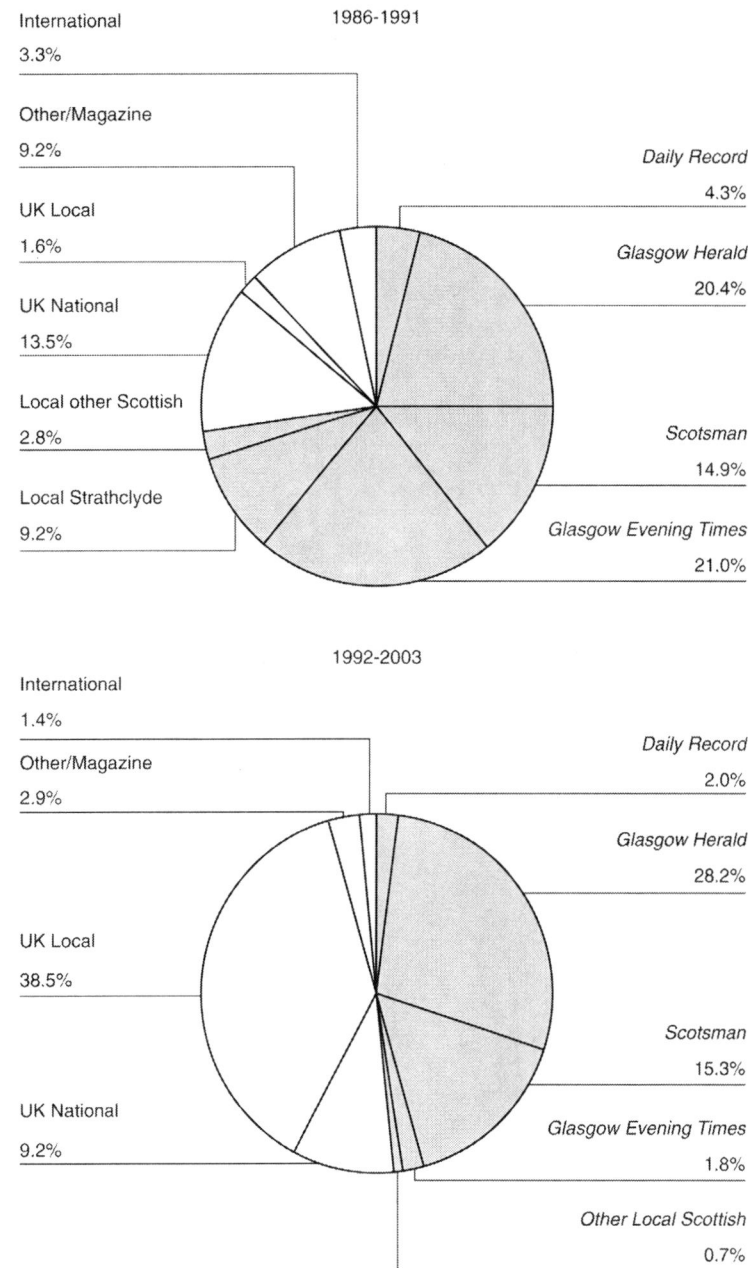

Figure 1. Provenance of newspapers analysed (percentages).

coverage and was thus the main source of
the narratives emerging from the first period
(72.6 per cent). The rest of the UK coverage
is marked by the predominance of national
broadsheets (13.5 per cent), most of which
have a Scottish base as well. In the post-1990
period, however, there is a greater balance
between Scottish-based and rest-of-the-UK
papers, with a marked predominance of UK
local dailies (38.5 per cent) followed by *The
Herald* (28.2 per cent). This implies that
post-1990 narratives are less dominated by
Glasgow-based papers and are instead more
likely to be influenced by opinion leaders
external to the city.

Moving into the narratives as such, in the
1990 period, eight thematic areas were identified, each of them sub-divided in sub-themes
including, in the case of outstanding levels of
coverage, the names of specific events or

individuals. This was the case of Peter
Brook's epic theatre piece 'Mahabharata'
(placed within the broader theme of 'Highlight Events') and Glasgow's District
Council leader at the time, Pat Lally (within
'Organisation, key figures and policy). When
analysing the period from 1992 to 2003, it
was necessary to adapt the emphasis as,
often, Glasgow 1990 was not the main
subject of the article but just a reference in
passing used to strengthen specific points. In
this context, four main thematic categories
were identified: discussion around the legacy
of 1990 on Glasgow's image or quality of
life; economic legacies; cultural legacies;
and governance or policy legacies. Table 1
shows the codes used for each period and
the way they interrelate.

Figures 2–5, together with Tables 2–4,
show the evolution of these thematic

Table 1. Coding table: main themes in 1986–91 and 1992–2003

1986–91	1992–2003
1 Representing Glasgow Quality of life Image and perceptions of Glasgow Promotion/city marketing	*1 Image legacies* Quality of life Image and promotions
2 Bringing business to Glasgow **Business and leisure tourism/visitor numbers** **Inward investment/job creation** **Infrastructural developments**	*2 Economic legacies* **Business and leisure tourism** **Economic and physical regeneration**
3 Physical legacies for culture Refurbishments/new cultural venues	*3 Cultural legacies* Cultural physical legacy
4 Performer and event origin Internationalism/parochialism Local talent/foreign imposition or elitism	Support to international artists/collaborations Support to local artists/cultural endeavours
5 Highlight events	**NA**
6 Money and funding Sponsorship and/or other private money State and/or council funding Ticketing and ticket sales	*4 Governance/policy legacies* Arts funding and finance NA
7 Event reach within Glasgow Accessibility for the people of Glasgow Accessibility for minorities within Glasgow Participation	Social inclusion and access to the arts
8 Organisation, key figures and policy Council leadership Event leadership Cultural policy developments	Governance, leadership and policy

Note: The bold within the table shows the transfer from one period to the next.

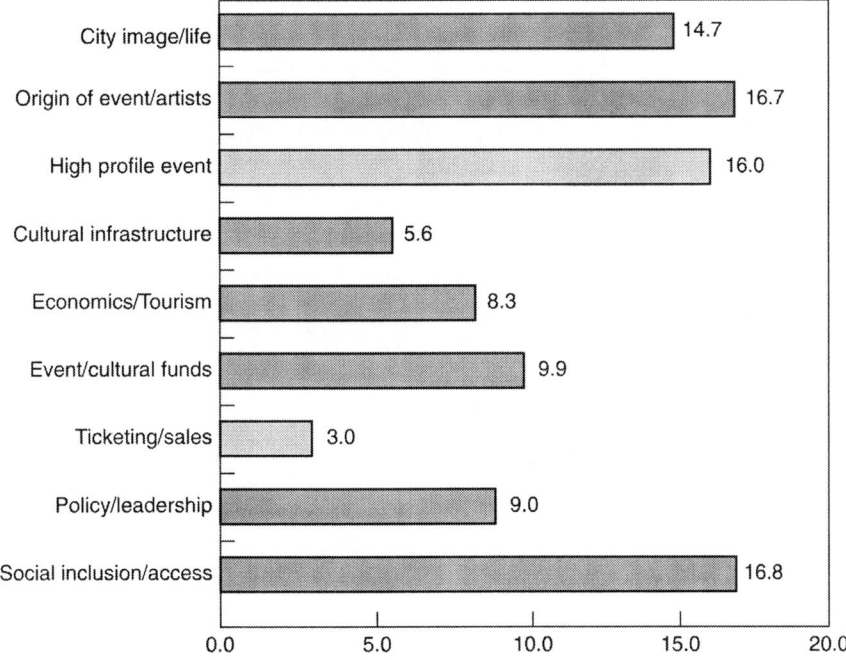

Figure 2. Distribution of main thematic areas, 1986–91. *Note:* the 'high profile event' and 'ticketing/sales' categories are shaded differently to indicate that they were related strictly to the ECOC programme and will not be compared with post-1990 coverage.

references and the attitudes adopted by the press when reporting them. In the period leading to 1990 and its immediate aftermath, the most frequent issues under discussion were the abilities of 1990 to enhance access to the arts and encourage direct participation. This was closely followed by the origin of events and performers, then discussion about the events included in the ECOC programme,[10] and Glasgow's image and quality of life.

The remarkable amount of discussion around access and inclusion indicates that the decision to present a programme that combined traditional arts with other activities of interest to non-arts audiences and minority

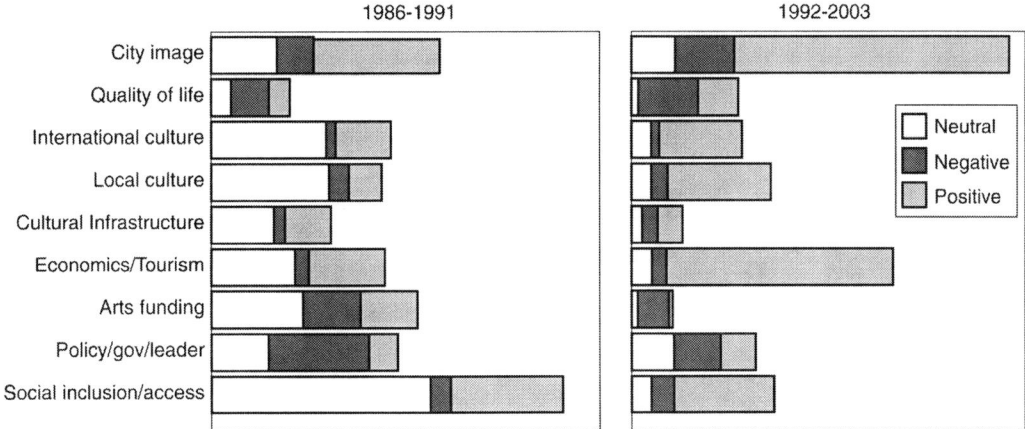

Figure 3. Thematic coverage and attitudes by theme, 1986–2003.

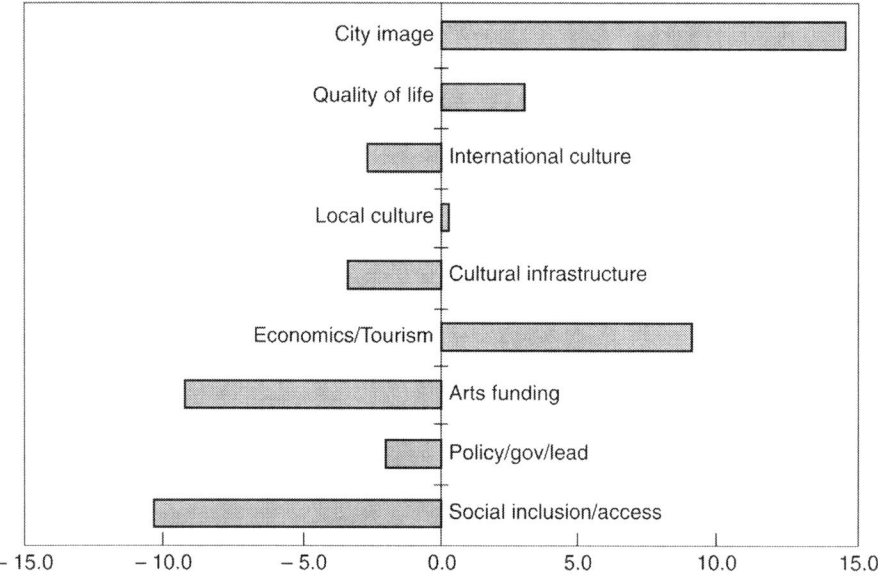

Figure 4. Thematic rate of change between 1986–91 and 1992–2003.

groups had a strong impact. Moreover, 32 per cent of these references were positive (with 62 per cent being neutral and only 6 per cent negative), which suggests that there was a high level of satisfaction among local communities and their opinion leaders. However, it is also important to note that up to 59 per cent of the coverage on access was presented through Glasgow city and regional tabloids—typically representing the interests of lower-income groups—rather than national Scottish or national UK papers—addressing the interests of middle-class readers. In this sense, it is possible to question the overall level of

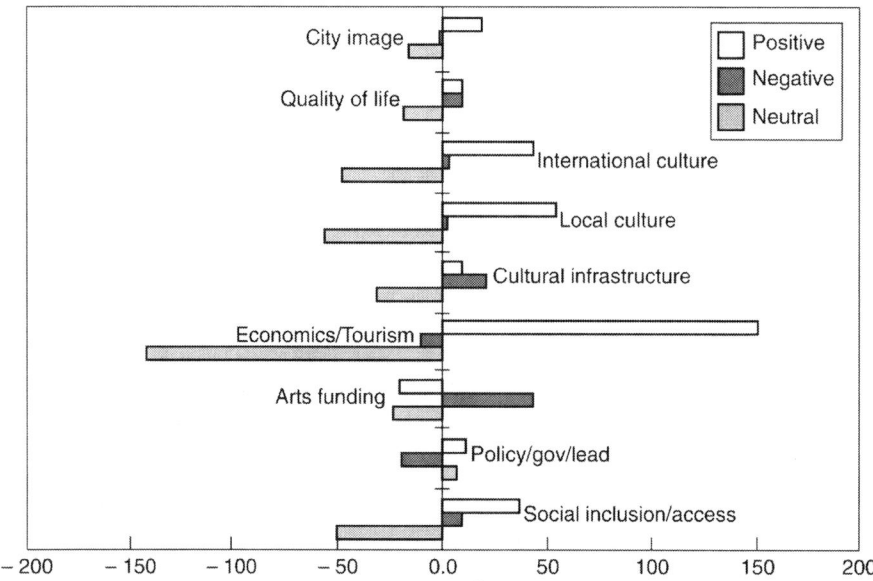

Figure 5. Attitudinal change by theme between 1986–91 and 1992–2003.

Table 2. Main thematic areas discussed, 1986–91

Attitudes/theme	Neutral	Negative	Posititive	Total
City image	27.9	16.8	55.3	100
Quality of life	23.8	48.4	27.9	100
International culture	64.3	4.9	30.8	100
Local culture	68.5	13.4	18.1	100
High-profile events	81.5	7.5	11.0	100
Cultural infrastructure	50.7	10.7	38.6	100
Economics/tourism	47.1	8.8	44.1	100
Event/cultural funds	42.6	31.3	26.0	100
Ticketing	19.2	60.6	20.2	100
Policy/leadership	29.2	54.7	16.1	100
Social inclusion/access	61.6	6.3	32.1	100

recognition of this aspect of the programme beyond those communities directly involved and the papers representing them. The apparent division between the reporting of locally based tabloids and national broadsheets hints at an unresolved conflict of interest between those Glasgow communities—in the city centre and outlying estates—characterised by low levels of income, poor education and a traditionally marginal role in arts and cultural activity, and the most privileged groups or city élites, characterised by high acquisition levels and a history of direct involvement in culture, the arts and a strong position within city politics and economic affairs.

This situation relates to the theme coded as 'Glasgow identity and quality of life' and is presented in Figure 2 and Table 2 under the overarching theme of 'City image/life'. Although the generic theme was quite prominent (14.7 per cent), the 'identity and quality of life' issue represents only 3.7 per cent,

while the other 10.8 per cent refers to Glasgow's image and marketing strategy. These two aspects (image and quality of life), despite being strongly interrelated, were discussed in almost oppositional terms (see Figures 3–5 and Tables 2–4). Thus, while references to Glasgow's image change tend to be positive (55.3 per cent, with only 16.8 per cent of negative references and the rest neutral), references to its quality of life are notably negative (48.4 per cent, contrasting with 27.8 per cent positive). As suggested earlier, most discussions around quality of life focused on the perceived divide between the renaissance of the city centre and the slow progress made in the city as a whole, including its disadvantaged areas. As in the case of references to community access, most references to this issue were concentrated in local tabloid newspapers such as the *Glasgow Evening Times* (19 per cent of all references) but they are

Table 3. Thematic rate of change between 1986–91 and 1992–2003 (percentages)

Themes	1986–91	1992–2003	Percentage change
City image	13.5	28.1	14.6
Quality of life	4.7	7.8	3.1
International culture	10.6	8.2	−2.4
Local culture	10.0	10.3	0.3
Cultural infrastructure	6.9	3.7	−3.2
Economics/tourism	10.2	19.3	9.1
Arts funding	12.2	3.0	−9.2
Policy/gov/lead	11.1	9.1	−1.9
Social inclusion/access	20.8	10.5	−10.3
Total	100	100	

Table 4. Attitudinal change by theme between 1986–91 and 1992–2003 (percentages)

Attitudes	Neutral	Negative	Positive
City image	−16.7	−0.8	17.5
Quality of life	−18.0	9.3	8.7
International culture	−48.0	4.2	43.7
Local culture	−55.5	1.1	54.3
Cultural infrastructure	−30.7	21.3	9.4
Economics/tourism	−142.0	−8.4	150.3
Arts funding	−22.6	43.7	−21.0
Policy/gov/lead	6.8	−18.6	11.8
Social inclusion/access	−48.7	10.9	37.9
Total	−375.4	62.68	312.7

also strongly reflected in UK national newspapers.

> Glasgow is more divided now than ever. It is two economies and you really can't have two cities. Everyone should feel enfranchised ... Some of the heat from the centre must percolate out to the schemes. As a city, we must get a sense of balance (Hague, 1989).

In any case, the clear predominance of positive references about the ECOC's accessibility indicates that the overall narrative around 1990 was supporting the view that the event was inclusive and succeeding in generating a buzz both in the city centre and outlying estates.

The overall narrative was also establishing that Glasgow was being successful in transforming its image. The 55.3 per cent of positive references—the highest level in all of the coded themes—is even higher when separating discussion around image and perceptions properly (7.1 per cent of all themes, presented in a positive light in 70.2 per cent of cases) from discussion on the city's marketing strategy (discussed in 3.7 per cent of clippings, with 26.7 per cent positive references while attracting 29.2 per cent negative coverage). The latter denotes that while the press agreed that 1990 had improved the perception of the city, there was a certain cynicism about Glasgow's aggressive approach to marketing. Most of the positive references in the marketing category refer to the successful 'Glasgow's Miles Better Campaign',

launched in 1983 and considered the "first step in ... a slow process of urban revival" (Paddison, 1993, p. 346). Negative references, on the other hand, concentrated on controversies around the chosen slogan for the ECOC ('In 1990, There's a lot Glasgowing on') criticised by its offering a sanitised version of the city and for being a sell-out to English audiences (see Boyle and Hughes, 1991, p. 225; McLay, 1990, p. 87; Mooney, 2004, p. 331).

Contrasting with the local focus of debates on community access, discussions around Glasgow's image occur more frequently in national and overseas papers. This was particularly the case during 1990, when this theme reached 12 per cent of all coverage (as opposed to an average 7.1 per cent between 1986 and 1991). However, while the average was 12 per cent, Scottish national papers dedicated only 9 per cent of their coverage to image issues, a percentage that reaches 22 per cent in UK national papers, 40 per cent in non-Scottish local papers and approximately 60 per cent in papers based abroad. This progression is also reflected in the level of positive attitudes, international papers being positive in 87 per cent of cases between 1986 and 1991. A representative selection of international headlines is indicative of the celebratory tone surrounding references to Glasgow's image transformation

'Glasgow's No Mean City Anymore' (*Wall Street Journal*, 14 January 1988).

'The City that washed its face, Glasgow— one-time Industrial Metropolis; European

City of Culture 1990' (*Der Tagesspiegel*, 29 January 1989).

'Newly scrubbed of soot, the city bustles with cultural events, commerce and fresh hope' (*New York Times*, 25 June 1989).

'Glasgow's reputation as Scotland's biggest, dirtiest, slummiest, most violent city is no more ... The ghost of an ugly past has been laid to rest' (*Sydney Morning Herald*, 13 July 1989).

'The ugly duckling of Europe has turned into a swan' (*Los Angeles Herald Examiner*, 27 August 1989).

'From tough industrial town to cultural mecca' (*Vancouver Sun*, 10 March 1990).

'Glasgow comes out of the black hole' (*Journal de Geneve*, 25 August 1990).

While it must be noted that these are headlines and, as such, they use an overemphatic tone, it can be argued that this sort of rhetoric had considerable influence on public opinion within the UK and Glasgow because it was often used as direct quotes within event and other tourist promotional materials such as newsletters and special brochures.

Discussion about the origin of events and artists was the second most dominant theme of the period (16.7 per cent). It is divided in two sub-categories that received similar levels of coverage: discussion on the levels of support to showcase local artists and levels of support to bring international stars. As show in Table 2, the first category attracted higher levels of criticism (13.4 per cent as opposed to 4.9 per cent), as many journalists felt that the main programme did not place enough emphasis on portraying home-grown talent. Alternatively, references to international stars were quite positive (30.8 per cent as opposed to 18.1 per cent) and emphasised their relevance to prove Glasgow's status as a European cultural centre.

The significance of other issues such as economic impacts, arts funding and city governance are illustrated in Figures 3–5 and Tables 3 and 4. Together, these permit an assessment of the evolution of references to 1990 throughout the following decade and

will help to establish which themes have sustained a currency in the long term and, consequently, which issues are seen as the real legacy of the event and on which grounds. In order to facilitate the comparison, both periods are presented under the same thematic categories and issues that were strictly related to 1990 (such as high profile events) have been dropped.

The study of the progression of references shows that, over the years, there has been a marked shift in the themes most strongly associated with 1990 (see Figure 3). Most notably, during 1992–2003 there has been continuous growth of references to the image of the city, while discussions around the impact of 1990 on access to the arts and social inclusion have decreased dramatically. At the same time, merely descriptive coverage has practically disappeared and been replaced instead by an overwhelming majority of positive coverage, with the notable exception of references to arts funding.[11] This is partly explained by the growing remoteness of the event. References to the ECOC after 1992 tend to occur in order to stress particular points which usually involve a clear positioning on the part of the journalist, rather than as a simple statement of fact or information update as it was the case in the lead to and during 1990 (for example, daily highlights). Figures 4 and 5, together with Tables 3 and 4, show rates of thematic and attitudinal change in percentages.

The themes that have become most strongly associated with Glasgow 1990 in the 1992–2003 period are the image of the city (+14.6 per cent), followed by economics and tourism (+9.1 per cent) and, at a lower level, references to Glasgow's quality of life (+3.1 per cent) (see Figure 4 and Table 3). In contrast, among the issues that lost media coverage are the ECOC's impact on access and participation in the arts (−10.3 per cent), followed by references to cultural funding (−9.2 per cent), cultural infrastructure (−3.2 per cent) and the internationalism of Glasgow's cultural scene (−2.4 per cent). Discussions around the effect of 1990 on Glasgow's home-grown talent and cultural abilities have remained fairly unchanged (a slight increase of 0.3 per cent).

The most remarkable improvement in attitudes (see Figure 5 and Table 4) relates to the impact of 1990 on attracting conference and leisure tourism and accelerating the city's wider economic regeneration (+150.3 per cent of positive references). This is also one of the few categories for which negative coverage has declined (−8.4 per cent). The progression demonstrates that tourism growth and economic development are two of the strongest arguments presented as evidence of ECOC success today.

Positive attitudes have also increased regarding the ECOC's impact on developing and sustaining support to home-grown cultural talent (+54.3 per cent) and bringing international artists to the city (+43.7 per cent). In the latter case, there has also been a slight increase of negative reporting (+4.2 per cent), which suggests that self-congratulatory remarks about local nurturing have been accompanied by accusations of 'parochialism' and excessive 'cultural protectionism' in Glasgow, both terms heavily used by journalists which have also emerged in our personal interviews with grassroots arts organisations, established artists and business leaders.

> Glasgow is a victim of its own propaganda. It believed it was a world-class city with a world-class economy. But in over a quarter of a century and after perhaps three billion pounds of public subsidy, not one world-beating company has emerged from the second city. ... The culprits in Glasgow's long decline are threefold. The dispirited middle classes who fled the city. The Pol Pot planners whose social engineering halved the city's population. And one-party city government—introverted, sectional, arrogant, parochial and incapable of appealing outside its own narrow constituency (Kerevan, 2000).

Despite the above criticisms, references to Glasgow's image transformation have been sustained at a very positive level, which reinforces the view that this remains a core source of arguments about success.

Overall, the analysis of press coverage shows that Glasgow's image, tourism and economic renaissance are presented as the main legacies of 1990, while access and participation in culture, arts funding, cultural infrastructure and internationalism have progressively diminished in current debates and thus lost currency as possible legacies. Furthermore, attitudinal changes indicate a growth of tolerance towards economic discourses. The substitution of social (access) concerns by economic interests suggests, on the one hand, that Glasgow 1990's social objectives have not been seen to be very sustainable and, on the other, that economic discourses have become more central—and often the top priority—within any debate on urban regeneration. As noted at the start of the paper, this is supported by the progressive change of emphasis (again, from a social into an economic focus) of both academic and practitioner-oriented literature on cultural policy and culture-led regeneration (see Bianchini, 1990; Greenhalgh, 1998; Kong, 2000).

An assessment of how these media narratives transpire within the personal accounts of Glasgow's local creative groups today helps to identify how the legacy of 1990 is lived and thus provide arguments about the most valuable cultural benefits of hosting the ECOC in the long term.

Sustaining Long-term Cultural Legacies through the ECOC: Impacts on Local Identities

As is the case for Bailey et al. (2004) our project provides evidence that the most interesting and complex cultural legacy of Glasgow 1990 is its effect on the identities of the local community. Our understanding of how 1990 has impacted on the lives of Glaswegians and the long-term cultural benefits of such experience is a result of deconstructing the media narratives presented above and contrasting them with the current (14 years on) interpretation by selected interest-groups. In the CCPR project, we focused on local creative groups, defined as those individuals (representing an institution or themselves) directly involved in the production or development of cultural meaning in Glasgow. Cultural meaning was understood in a similar fashion to that associated with current definitions of the creative/cultural industries in the UK (CITF, 2001; Cunningham, 2002; Volkerling, 2001) and included the arts, the media and other fields of particular relevance to Glasgow such as design, fashion and architecture. Table 5 shows a summary of the cultural and creative fields represented by the participants in our interviews and focus groups.

For individuals working in the voluntary arts sector, grassroots cultural organisations and community arts groups, 1990 is seen as a critical point of reference that gave them the level of confidence they needed to keep working despite regular funding cuts and diminishing institutional support. They feel they learnt to be more entrepreneurial and claim that part of the reason why they have survived to this day is the memory of what took place in 1990, when they had access to mainstream funds and, on occasion, mainstream venues.

> We never returned to the baseline that we had before 1990 ... partly because we'd learned a lot about funding, and putting together packages. ... We had been very dependent on one or two sources before that ... 1990 forced us to start looking wider, ... looking at the private sector, ... trust funds. ... We learned a lot ... We were forced to. It was a very painful process, but probably a good process in the end (community arts focus group, 16 September 2003).

In contrast, their disappointment concentrates around structural and policy issues as they believe that 1990 did not have a direct positive legacy on cultural policy and governance

> [Ten years later i]t's the immediate thing [that matters] and there is of course a major hang-up about meeting targets.

Table 5. Interviewees and focus group participants: thematic categories

Categories/interest areas	Number of interviewees/focus groups participating
Local authorities, public agencies and quangos	*Total*: 17 interviews; 1 focus
City/district council, regional authorities (1986–2003)	Interviews: 7 city, 5 region; 1 focus group (city)
Development agencies, tourist boards	Interviews: 3 development agencies, 2 tourist board
Opinion leaders	*Total*: 16 interviews; 1 focus
Journalists covering the event and related city issues	3 interviews, 1 focus group
Academics researching the event/city regeneration	11 interviews/informal meetings
Local activists/city writers	2 interviews
Creative groups	*Total:* 12 interviews; 4 focus; survey 30
Grassroots arts organisations	2 interviews, 1 focus group
Established artists and cultural venues representing: painting, sculpture, theatre, alternative performance arts	3 interviews, 2 focus groups (visual artists/ galleries, venue-based institutions)
Young artists (GSA graduates)	1 survey of 30 young people
Creative industries—excluding traditional arts (i.e. design centres, architecture centres, music studios, film studios)	1 focus group
Event organisers (producers, programmers, PR)	6 interviews
Overall total	45 interviews; 6 focus groups; survey of 30 people

Note: See further details about interviewee, focus group and survey profiles in Appendix 1.

You've got to have *x* number of people through your door, or your box office. ... It's actually got nothing to do with ... [the] quality of people's life, or how they engage ... in a creative experience. ... The quality of what we do is not really high on the agenda, so ... then the policy will not be high on the agenda [either]. [The local authorities and funders a]re not looking at what to change [in cultural policy terms] [and do not] have what ... used to be called the 'social impact of the arts'. I don't think that exists anymore (community arts focus group, 16 September 2003).

This view is shared by representatives from traditional art-forms (performing and visual artists) and established cultural venues. However, grassroots arts groups and established artists/venues tend to have opposite views about the direction that urban cultural policies should take today. Grassroots organisations lament the lack of effort to sustain 1990 community initiatives and structures in the event's immediate aftermath. They remark that local government reorganisation in the mid 1990s accentuated the problem by further diminishing funding for the arts and provoking a radical change of orientation in local policy (see García, 2004a).[12] But they tend to view the near future with optimism as they see the current City Council emphasis on social inclusion (see GCC, 2001) and the expanding cultural interests of housing associations and other institutions in Social Inclusion Partnership (SIP) areas as an opportunity to argue the case for the arts as a catalyst for social inclusion and multicultural understanding and regain some protagonism in the city and outlying estates.

[Our work in North Glasgow and Easterhouse] ... hasn't been easy, it's a slog like everything else, but you know there are people willing to sit down with you from development agencies and helping you write your funding applications. There [is] support from all these different aspects, ... and I think it's got a lot to do with the SIPs, love them or hate

them—and I [do] both. I think ... they are driving it. [They are claiming t]hat there should be organisations, arts organisations, resident in specific areas and I think it's certainly quite a cushy position to be in some areas, servicing a particular SIP area because there is such a huge support network. And then from larger arts organisations there are things developing, like Fablevision is working with us. Yeah, and it felt quite strange to realise. It did take a year or so of digging around for the penny to drop, but when it did, yeah, there is so much of a resource there, its really nice (community arts focus group, 16 September 2003).

In contrast, business leaders, established artists and cultural venues complain about Glasgow's loss of profile as a centre of excellence, the reduction of links with overseas artists and major events and what they call the growing "parochialism", "protectionism" (business focus group, 24 September 2003) and "excessive populist emphasis" of the city leaders (visual arts focus group, 30 September 2003).

[*Participant 1*] [We need to ensure that] all the relevant stakeholders [in events coming to Glasgow, etc.] can see what the plan is, what's happening, what's on the SECC [Scottish Exhibition and Conference Centre] diary, you know, one easy to read process.

[*P2*] I sincerely hope so, but I suspect that there is going to be clashes between what the events manager of Glasgow [City Council] wants and what the new chief executive of [the new quango] EventScotland ... wants. We have had several near misses [in past event bids].

[*P1*] So if everything is brought together in one handy to read form, then Scotland can dip into that and yield the country.

[*P3*] Yes that could be done if the people concerned got away from parochialism, jobs-worth, large salaries and feeling they have to do something to justify it.

Everything is protectionism not partnership (business/tourism focus group, 24 September 2003).

Some artists feel that 1990 showed Glasgow at its best because the event's artistic directors were daring and provocative and the authorities allowed them to go ahead with their views and, most importantly, funded them. This has had an impact on the local artists' "vision" and "aspirations" but, according to them, has been constrained since by the "conservatism" of current funding bodies and has resulted in artistic migrations to other parts of the UK or Europe (cultural venues focus group, 6 October 2003)

> [*Participant 1*] I don't think there is a critical mass at the moment. I think it is very feeble and people are trying to pull it together but it just doesn't seem to be happening. Which is to do with all these other factors, lack of resources, lack of money, and political will and the uses of art. ... And the less that happens the more people will ... disappear. They just move elsewhere.

> [*P2*] When you are approached by a company or an artist ... there is a kind of tunnel-vision approach. They are not really, really going for it. ... It is a very conservative approach to how they want to do this show and it is basically [that] the parameters are already set and they won't go beyond them. And it is very kind of 'we've got this show in front of us, that's what we are going to concentrate on'. Rather than, 'I'm director of this company, and my vision extends beyond this show'. ... It all feels very, very isolationist, and everyone is in their own little pockets and they are just trying to get on with it (cultural venues focus group, 6 October 2003).

Eight out of ten established artists and cultural venue managers argued that the ECOC was important for the city and for their own personal development but that without a political environment sensitive to avant-garde artistic endeavour, the memory of it is also a source of frustration for "what was possible [back in 1990] and is not [happening] anymore" (cultural venues focus group, 6 October 2003).

According to creative entrepreneurs in the film, television, music and design industries, 1990 was an initiative that helped to regenerate the city's economy but had a low impact on the city's cultural production scene. They resent that the event focused on consumption rather than production, which has led them to question its sustainability. In this sense, they doubt that the ECOC has been a catalyst for the city's cultural industries and do not feel that the event has had a direct impact on their work. This contradicts the official version of Glasgow development agencies, who claim that 1990 provided the necessary arguments to establish a city-specific strategy to support the creative industries, an initiative currently under discussion (see EKOS, 2004). During a personal interview, the director of Regeneration Services at Glasgow City Council tried to bridge this apparent contradiction. He did so by summarising the main findings of a report produced by the think-tank Comedia on Glasgow's creative assets post-1990 (Comedia, 1991) and referring to the Council's response to the report recommendations

> [The report indicated that] 1990 had been all about consumption: people came, they visited events, they spent money, and ... they participated in things which they might not have participated in. But the benefit was largely visitor expenditure, community development [and] community participation; there was no real production feel to it. What [the Comedia report did was] ... split Glasgow's creative economy into the performing arts, the visual arts, music, design and architecture, ... fashion. And [it indicated that] in each of those areas there is a consumption part to the economy but there is also a huge production industry, or potential for production. ... Now I don't know how much the original commissioners paid for that piece of work, but they took it and put it on a shelf. We [the City Council] took it

and used it. So . . . I did a report . . . that said we need to make more of what we invested in . . . 1990 . . . to seek more lasting benefits; and we do that by doing production-based activities as well as events. . . . So we had a music business development programme . . . We also established the Glasgow Film Office. . . . [And] we did a lot on architecture and design between 1996 and 2001 . . . We still do a lot on the consumption side [as well], we are in the process of developing a new events strategy . . . and developing cultural tourism, which features quite highly in the marketing of Glasgow (personal communication, Glasgow City Council Regeneration Services, 25 September 2003).

Despite these explanations, the sustained criticisms emerging from the focus groups suggest that Glasgow's creative groups are sceptical of the official version presented by local authorities and quangos about Glasgow's success in culture-led regeneration. Creative entrepreneurs maintain that, if an event as generously funded and publicly acclaimed as Glasgow's ECOC has not resulted in clearly sustainable cultural schemes, there is little hope for any other such initiative to change the trend. They see more value in direct support to individuals to allow "organic growth" than what they view as an "obsession" with a "carnival of grandiose schemes" generating "massive advertising kudos" perhaps but divorced from the reality of the city and its "creative soul" (creative entrepreneurs focus group, 15 January 2004).

Younger generations, contacted through a personal survey of 30 graduates from the Glasgow School of Art (GSA) in June 2003, do not see 1990 as a reference-point. However, their reasons for being attracted to Glasgow offer some insight into the factors that are most valued in the city (see Table 6). The existence of a strong arts scene, the overall vibrancy of the city and the strong sense of community resulting from its compactness are among the most valued factors. This suggests that young

Table 6. Most valued factors of Glasgow city life for GSA graduates ($N = 30$)

Art scene/community	11
'Lots going on'	8
Humour/friendliness	6
Home town	6
Pubs/nightlife	5
Compact city	4
Near to the Highlands	4
Shopping	3
Affordable	3
Big city	2
Music scene	2
Architecture/looks	2
Industrial/working class	2
Theatre	1

graduates agree with the rest of interviewees and focus group participants that Glasgow offers a good environment for creative development. A common expression is that the city is small enough to give a "sense of community" and diverse enough to feel "cosmopolitan" and "trendy" (arts graduates survey, June 2003; creative entrepreneurs focus group, 15 January 2004).

At this point, the remaining question is whether 1990 is seen as an important contributor to the city's current vibrancy. Representatives from the business world, including tourism agencies, shopping centres, conference organisers and hotel managers, supported the view that the ECOC boosted Glasgow's attractiveness as a centre for business and leisure.[13] They saw this as a cultural as well as an economic legacy. In contrast, the arts community (in particular, visual artists, writers and musicians) argued that Glasgow's cultural vibrancy had been there well before 1990 and was the result of a strongly rooted sense of community and a tradition of activism and civic involvement.

Glasgow was before and is now still one of the most . . . culturally committed cities in the UK without a shred of a doubt. . . . The Year of Culture definitely put Glasgow on the map as a European city. [But] maybe you could argue it's always been a European city, it's just never been acknowledged (Personal communication,

Glasgow-based theatre company and visual arts venue, 17 October 2003).

In the cultural venues focus group, a further reason for such a cultural vibrancy was a strong leadership that, in their view, has weakened over the years

[*P1*] [We can say there is a Glasgow model of cultural regeneration] to certain extent, but you have to have the creativity within the city at that point. And you can't really manufacture that. The Glasgow model worked because there were the communities and creativity there.

[*P2*] Plus the leadership.

[P1] Plus the leadership as well. All these things come together. But if you have the money and you have the leadership, you still wouldn't be able to successfully develop that as a model, I don't believe, without having those creative communities there at that time (cultural venues focus group, 6 October 2003).

For 7 out of 10 artists and cultural venue representatives, the ECOC is seen as a sort of package used to sell this vibrancy to others. They recognise that this 'selling' strategy had attracted not only businesses and tourists but also other artists from around the UK and overseas, which, ultimately, added to the city's creative capacities. But they felt that the distribution of efforts between "championing the arts and championing city marketing" should have been more balanced, especially post-1990 (journalist focus group, 15 September 2003).

Overall, practically all of the individuals interviewed (9 out of 10) agree that 1990 played a role in Glasgow's renaissance and has become a point of reference for artistic and creative endeavour. However, the effect and benefit of such 'renaissance' (and with it, its impact on personal lives) are interpreted in different terms. The strongest differences are found between grassroots organisations and established artists and venues. The former feel that, despite the event's poor structural legacy, the confidence they gained

in 1990 is allowing them to influence the current approach of local authorities and quangos to social inclusion. The latter are, however, disenchanted with the lack of leadership and aspirations of local élites and feel that Glasgow's creative potential has only survived thanks to the strength of the city's informal art community networks. Beyond this, all creative groups with the exception of business/leisure representatives, coincide in their scepticism about the value of the event's economic legacies, a perspective that clearly opposes the main narrative emerging from recent press coverage. This suggests that there is a marked divide between the external and internal image legacy of Glasgow 1990, a divide that is in fact reflected by the growing difference in emphasis and attitudes towards city issues in Glasgow household papers as opposed to UK national or international papers.

Conclusions. Long-term Benefits: Beyond Economics?

The analysis of media and personal narratives on Glasgow 1990 more than a decade later show that the event secured some important long-term cultural benefits. The findings suggest that it is the softer, less tangible cultural benefits that have been better sustained, while other widely acclaimed economic benefits such as job creation are questioned both by local creative groups and recent academic publications (see Turok and Bailey, 2004, p. 169). This is because, although it cannot be denied that 1990 did contribute to job creation in the service sector (for example, with the fillip it gave to the tourist industry), the quality of the jobs was often relatively poor and rarely provided the transferable skills that people need to remain in the job market in the long term. Furthermore, it is difficult to disentangle the specific long-term employment effects of 1990 from other developments that emerged in the decade to follow.

Overall, the contrast between the long-term survival of memories linked to creative

personal development and the poor maintenance (or local appreciation) of tangible outcomes—such as the establishment of
creative production centres and related structures to secure cultural provision—indicates
that hosting an ECOC can lead to a marked
imbalance between the sustainability of tangible and intangible benefits. Yet, as in the case
of economic impacts, it is also worth noting
that the complex nature of intangible cultural
legacies makes it difficult to conclude
whether they are a direct result of a particular
event or culture-led regeneration strategy. In
Glasgow, the most valuable cultural legacies
interrelate with other elements that are
inherent to the fabric of the city and result
from many dimensions beyond 1990.

The difficulty of demonstrating the direct
impacts of regeneration will always create
temptation to conflate cultural with economic
or physical assessments in order to claim
'success'. However, as reflected in the study
of personal narratives, evidence of tourism
growth and office relocation is not proof of
improvement in local citizens' well-being.
Furthermore, as noted by Bianchini, accepting a business rhetoric to justify cultural
achievement may lead to other undesirable
situations

> Now there is the risk that the incorporation
> of the arts into urban growth coalitions will
> reduce the freedom which is necessary to
> perform this essential critical role. The
> hegemonic status of the belief that 'what's
> good for business is good for the city'
> could seriously weaken the ability of the
> arts to point at alternative notions of 'the
> good' for both the individual and the com-
> munity (Bianchini, 1990, p. 240).

Further, while it is common for city boosters
to rely on media accounts as a measure of
success, a trend evidenced by the heavy
reliance on press quotes to promote new or
commemorate past events,[14] it is only
through contrasting them with personal
accounts that we can claim an understanding
of cultural legacies. Thus, while retrospective
media content analysis provides a good basis
for understanding discourses contemporary to

past events, assessing their impact on personal
lives allows us to put the media hype in perspective. The study of personal narratives
also helps to contextualise the strong criticisms
emerging from academia and activist groups
opposed to event-led regeneration on the
basis of Marxist urban sociology and political
economy. The latter is explored by Stevenson
when discussing her own understanding of
"the politics of urban difference"

> A cultural approach to the idea of difference
> makes it possible to appreciate that, at the
> level of lived experience, spaces such as
> the home and suburbia which, when seen
> in structural terms are said to contribute to
> the subordination of certain groups, may
> at the level of the micro be sites of empow-
> erment for those groups. Such insights
> emerge most sharply from qualitative
> research that involves actually listening to
> the experiences and priorities of the
> people who use particular spaces, rather
> than focusing on overarching objective
> structures, such as class of patriarchy
> (Stevenson, 2003, p. 42).

Although Stevenson also warns against the
"dangers in focusing exclusively on micro-
situations" and the "idealization of city life"
(p. 43) the value of in-depth interviewing
and personal narrative cannot be undermined,
or denied the possibility of being used as evidence of positive or beneficial cultural legacy.
The only note of caution about the approach
taken here is that the analysis of narratives
by creative groups is not representative of
Glasgow as a whole. They speak from the perspective of cultural production and/or development rather than that of average citizens.
This was the approach chosen in order to
allow for greater consistency and hence to
gain a deeper understanding of emerging
issues, but it is not aimed at answering the
question of how 1990 has affected the city's
general population. This is an area that could
be further developed through other emerging
methodologies such as Rogerson's (1999)
approach to assessing quality of life in cities.

Lessons for 'ECOC-led' Regeneration ...

After almost 20 years, the ECOC programme could be seen as a mature initiative and a source of lessons to guide urban regeneration. The existence of internationally recognised 'success-stories' such as Glasgow has enhanced the prestige of the programme and generated growing expectations in cities aspiring to improve their image and boost their tourist economy. It may be no coincidence that Glasgow's most acclaimed legacies are also the two aspirations featuring most prominently in the aims and objectives of a majority of cities having since hosted the title (Palmer/Rae Associates, 2004, pp. 43–46).

However, despite its apparently good reputation, it is misleading to suggest that the ECOC offers a good strategic and operational basis for culture-led regeneration. This is mainly because of the poor standards of event monitoring and evaluation, particularly in the long term. This may change in the near future as, due to its ability to stimulate competition between cities, the programme is currently seen as a good contributor towards strengthening the European economy. To maximise the potential of the ECOC, some of the policy developments forecasted by Myerscough (1994, p. 5) are currently taking place. These include strengthening links with the Committee of the Regions and the European Parliament, and increasing the interaction with European Commission competences beyond the cultural remit, such as tourism, urban regeneration and training. But this approach brings again a possible limitation: ECOC's cultural regeneration is likely to be measured and justified in non-cultural terms.

The research presented here and the work by Bailey *et al.* (2004) emphasise that it is possible to assess cultural impacts and legacies, that the process must be longitudinal and that this exercise is fundamental to gain a full understanding of the effects of culture-led regeneration. Without denying the value of determining economic, physical and social impacts, the main argument is that culture needs to get back to the centre of any discussion on this topic. Otherwise, programmes such as the ECOC may become meaningless and easily dispensable. If the core objective is attracting tourism rather than enhancing the city's artistic and cultural life, hosting the Capital of Culture could be easily replaced by large business conventions, global sport competitions or any major corporate event, without mattering whether these events are sensitive or not to the character and cultural roots of their local hosts.

The narratives and experiences recalled in this paper should be seen as a valuable indicator of success in regeneration and as a first step towards establishing credible and replicable approaches to evaluating cultural impacts. In this sense, any advancement into cultural impact assessment must begin with a broader conceptual and methodological framework regarding the acceptance of 'evidence' than is the case today. Funding bodies and researchers in this field should acknowledge that the assessment of discourses about a place and its people over an extended period of time is valid evidence. Ultimately, through studying Glasgow's experience, this paper claims that further developing techniques to understand cultural impacts seems the most feasible way to ensure the survival of the ECOC as a meaningful, effective and sustainable example of culture-led regeneration.

Notes

1. The other two models are termed 'cultural regeneration' and 'culture and regeneration'. According to the authors, the main difference is that while in the 'culture-led regeneration' model cultural activity is seen as the catalyst and engine of regeneration—the case of high-profile events such as the ECOC—in the 'cultural regeneration' model, "cultural activity is integrated into an area strategy alongside other activities in the environmental, social and economic sphere"—as is the case within cultural planning. Finally, in a 'culture *and* regeneration' model, "cultural activity is not fully integrated at the strategic development or master planning stage" and the intervention is often small (Evans and Shaw, 2004, p. 5).

2. For example, in referring to common tests to measure the environmental impacts or

cultural regeneration, Evans and Shaw (2004, p. 6) refer to "Quality of Life (ODPM's *local quality of life* indicators), Design Quality Indicators (DQI—CABE/CIC), Re-use of brownfield land"; and in the area of economic impacts, they mention "Employment/unemployment rates, income/spending and wealth in an area, and distribution by social group and location, employer location, public–private leverage".

3. This situation has been noted in the study by Palmer/Rae, which recommends the introduction of competitive processes of selection as a standard practice within respective countries as one of the key mechanisms to strengthen the relevance and public impact of future ECOCs (Palmer/Rae Associates, 2004, pp. 173–174).

4. The levels of positive coverage from Glasgow's nomination in 1986 to 1991 were much lower than in the post-event period (29 per cent positive, 19 per cent negative and 52 per cent neutral). In the period from 1992 to 2003 (which includes the coverage of the bidding process for 2008), positive references grew up to 61 per cent, followed by 21 per cent negative references and 14 per cent neutral. This is analysed in detail later in the paper.

5. Harvey argues that prestige arts-led and related large-event regeneration projects function as a "carnival mask that diverts and entertains, leaving the social problems that lie behind the mask unseen and uncared for" (Harvey, 1989, p. 21). He adds that "The formula smacks of a constructed fetishism, in which every aesthetic power of illusion and image is mobilised to mask the intensifying class, racial and ethnic polarisations going on underneath" (p. 21).

6. These indicators are partly based on the 'social benefit indicators' resulting from the report *Creating social capital: a study of the long-term benefits from community based arts funding* (Williams, 1996) as cited by Evans and Shaw (2004, p. 30).

7. See Table 5 and Appendix 1 for an indication of the profiles of interviewees and focus group participants.

8. Original quote

   Mis consideraciones acerca de los medios de comunicación enfatizan el papel que éstos desempeñan en los cambios culturales y la envergadura antropológica de los cambios producidos por la comunicación. Es decir, tal y como indica el título de mi libro, 'De los medios a las mediaciones', intento pensar no sólo los medios sino también los fines: cómo están cambiando

   los modos de constitución y reconocimiento de las identidades colectivas y la incidencia en la reconstitución de éstas tanto de los medios como de los procesos de comunicación (Martín-Barbero, 1991, p. 4).

9. See also Appendix 2 for a brief summary of our approach to press content analysis.

10. Note that general preview and review articles were excluded from the analysis. Only articles discussing particular events or the overall programme in the context of the ECOC and/or Glasgow's cultural regeneration are included here.

11. The remarkable increase in negative references to arts funding (see also Figure 5 and Table 4) denotes a disappointment in terms of the ECOC's ability to strengthen the economic position of arts activities in Glasgow. This was also a common complaint within the interviews and focus groups, as exposed in the next section.

12. Local government reorganisation was a Scotland-wide process that took place in 1996. In Glasgow, this meant the disappearance of the Strathclyde Regional Council, a key player in 1990 that, combining priorities and resources with Glasgow District Council, made possible the acclaimed balance in cultural provision—élite and grassroots—and spatial distribution—centre and periphery—so unique to Glasgow's celebrations. The changes in local government resulted in a break with emerging cultural policies born out of the 1990 experience (García, 2004a, p. 320).

13. See the profile of interest groups represented in Appendix 1.

14. This trend is also apparent in the proliferation of press clipping agencies contracted by local authorities and/or private event managers as a basis for up-to-date assessments of public opinion.

References

AHUVIA, A. (2001) Traditional, interpretive, and reception based content analyses: improving the ability of content analysis to address issues of pragmatic and theoretical concern, *Social Indicators Research*, 54(2), pp. 139–172.

BAILEY, C., MILES, S. and STARK, P. (2004) Culture led urban regeneration and the revitalization of rooted identities in Newcastle, Gateshead and the North East of England, *International Journal of Cultural Policy*, 10(1), pp. 47–66.

BERELSON, B. (1952) *Content Analysis in Communication Research*. Glencoe, IL: Free Press.

BIANCHINI, F. (1990) Urban renaissance? The arts and the urban regeneration process, in: S. MACGREGOR and B. PIMLOTT (Eds) *Tackling the Inner Cities: The 1980s Reviewed, Prospects for the 1990s*, pp. 215–250. Oxford: Clarendon Press.

BIANCHINI, F. (1999) Cultural planning for urban sustainability, in: L. NYSTRÖM and C. FUDGE (Eds) *Culture and Cities: Cultural Processes and Urban Sustainability*, pp. 34–51. Stockholm: The Swedish Urban Development Council.

BOOTH, P. (1996) *The role of events in Glasgow's urban regeneration*. Strathclyde Papers on Planning No. 30. University of Strathclyde, Glasgow.

BOYLE, M. (1997) Civic boosterism in the politics of local economic development: 'institutional positions' and 'strategic orientations' in the consumption of hallmark events, *Environment and Planning A*, 29(11), pp. 1975–1997.

BOYLE, M. and HUGHES, G. (1991) The politics of representation of the 'real': discourses from the left on Glasgow's role as European City of Culture, *Area*, 23(3), pp. 217–228.

BOYLE, M. and HUGHES, G. (1994) The politics of urban entrepreneurialism in Glasgow, *Geoforum*, 25(4), pp. 453–470.

BRYMAN, A. (2001) *Social Research Methods*, pp. 177–192. Oxford: Oxford University Press.

CITF (2001) *Creative industries mapping document*. Department for Culture, Media and Sport (http://www.culture.gov.uk/creative/mapping.html).

CLARK, B. (2002) Flushing out the Scottish financial mafia: the shady case of the Glasgow Development Agency, *Spunk Org* (http://www.spunk.org/texts/groups/sfa/sp000503.txt).

COGLIANDRO, G. (2001) *European Cities of Culture for the year 2000: a wealth of urban cultures for celebrating the turn of the century*. Association of the European Cities of Culture of the year 2000, AECC/AVEC, Strasbourg.

COMEDIA (1991) *Making the most of Glasgow's cultural assets: the creative city and its cultural economy*. (Report to Glasgow District Council).

CUNNINGHAM, S. (2002) *From cultural to creative industries*. Paper presented to the *Second International Conference on Cultural Policy Research, 'Cultural Sites, Cultural Theory, Cultural Policy'*. Wellington, New Zealand.

DAVIES, P. and RUSSELL, M.-T. H. (2001) *Comparative analysis of time-limited cultural development projects including festivals and other Capital of Culture projects*. International Centre for Cultural and Heritage Studies, University of Newcastle.

DCMS (DEPARTMENT OF CULTURE, MEDIA AND SPORT) (2004) *Culture at the heart of regeneration*. Consultation Report, DCMS, London.

EGAN, J. (2004) *Skills for sustainable communities*. Report to the Office of the Deputy Prime Minister. London: RIBA Enterprises Ltd.

EKOS (2004) *Glasgow creative industries strategy*. Discussion document prepared for Scottish Enterprise Glasgow and Glasgow City Council.

EVANS, G. (2003) Hard-branding the Cultural City: from Prado to Prada, *International Journal of Urban and Regional Research*, 27(2), pp. 417–440.

EVANS, G. and FOORD, J. (2003) Shaping the cultural landscape: local regeneration effects, in: M. MILES and T. HALL (Eds) *Urban Futures: Critical Commentaries on Shaping the City*, pp. 167–181. London: Routledge.

EVANS, G. and SHAW, P. (2004) *The contribution of culture to regeneration in the UK: a review of evidence*. London: Department for Culture, Media and Sport.

FREY, H. (1999) *Designing the City: Towards a More Sustainable Urban Form*. London: E & FN Spon.

GARCÍA, B. (2004a) Cultural policy and urban regeneration in western European cities: lessons from experience, prospects for the future, *Local Economy*, 19(4), pp. 312–326.

GARCÍA, B. (2004b) Urban regeneration, arts programming and major events: Glasgow 1990, Sydney 2000 and Barcelona 2004, *International Journal of Cultural Policy*, 10(1), pp. 103–118.

GCC (GLASGOW CITY COUNCIL) (2001) *Best value review of arts and cultural events*. Cultural and Leisure Services, Glasgow City Council.

GREENHALGH, L. (1998) From arts policy to creative economy, *Media International Australia Incorporating Culture and Policy*, May(87), pp. 84–95.

GRIFFITHS, R., BASSET, K. and SMITH, I. (1999) Cultural policy and the cultural economy of Bristol, *Local Economy*, 14(3), pp. 257–264.

GULLIVER, S. (2002) *The cultural impact of a city: Glasgow*. Paper presented at the *Meeting of the English Core Cities*, Liverpool.

HAGUE, H. (1989) City's rebirth leaves poor on the margins, *The Independent*, 4 April.

HARVEY, D. (1989) *The Urban Experience*. Baltimore, MD: The Johns Hopkins University Press.

KEMP, D. (1990) *Glasgow 1990: The True Story Behind the Hype*. Gartocharn: Famedram, ArtWork.

KEREVAN, G. (2000) The sick city, *Sunday Herald*, 12 March, p. 9.

KIRKPATRICK, J. (1996) Design as a tool for cultural change: Glasgow's experience, in: J. VERWIJNEN and P. LEHTOVUORI (Eds)

Managing Urban Change, pp. 52–64. Helsinki: University of Art and Design Helsinki Uiah.

KONG, L. (2000) Culture, economy, policy: trends and developments, *Geoforum*, 31(4), pp. 385–390.

MACDONALD, S. (2002) Defining a decade: Glasgow and the 1990 City of Culture, *Locum Destination Review*, Autumn(9), pp. 50–52.

MACLEOD, G. (2002) From urban entrepreneurialism to a "revanchist city"? On the spatial injustices of Glasgow's renaissance, *Antipode*, 34(3), pp. 608–629.

MARTÍN-BARBERO, J. (1991) Dinámicas urbanas de la cultura, in: INSTITUTO COLOMBIANO DE CULTURA (Ed.) *Gaceta de Colcultura*, December (12), pp. 4–16.

MAY, T. (1997) *Social Research: Issues, Methods and Process*. Buckingham: Open University Press.

MCLAY, F. (Ed.) (1988) *Workers City: The Real Glasgow Stands Up*. Glasgow: Clydeside Press.

MCLAY, F. (Ed.) (1990) *The Reckoning: Public Loss, Private Gain* (*Beyond the Culture City Rip Off*). Glasgow: Clydeside Press.

MOMMAAS, H. (2004) Cultural clusters and the post-industrial city: towards the remapping of urban cultural policy, *Urban Studies*, 41(3), pp. 507–532.

MOONEY, G. (2004) Cultural policy as urban transformation? Critical reflections on Glasgow, European City of Culture 1990, *Local Economy*, 19(4), pp. 327–340.

MOONEY, G. and DANSON, M. (1997) Beyond "culture city": Glasgow as a dual city, in: N. JEWSON and S. MACGREGOR (Eds) *Transforming Cities: Contested Governance and New Spatial Divisions*, pp. 73–86. London: Routledge.

MYERSCOUGH, J. (1991) *Monitoring Glasgow 1990*. Report prepared for Glasgow City Council, Strathclyde Regional Council and Scottish Enterprise.

MYERSCOUGH, J. (1994) *European Cities of Culture and cultural months*. Glasgow: The Network of Cultural Cities of Europe.

PADDISON, R. (1993) City marketing, image reconstruction and urban regeneration, *Urban Studies*, 30(2), pp. 339–350.

PALMER/RAE ASSOCIATES (2004) *Study on European Cities and Capitals of Culture 1995–2004* (Part I). Brussels: European Commission (http://www.palmer-rae.com/culturalcapitals.htm).

PEACOCK, A. and RIZZO, I. (Eds) (1994) *Cultural Economics and Cultural Policies*. Boston, MA: Kluwer Academic Publishers.

RAE, J. H. (Ed.) (1993) Glasgow: a city of change, *The Planner*, June.

REASON, M. and GARCÍA, B. (2003) *Approaches to the newspaper archive: content analysis and press coverage of Glasgow's Year of Culture*. Unpublished working paper available at the Centre for Cultural Policy Research, University of Glasgow.

RICHARDS, A. (2002) Culture as circus, *Here And Now* (Glasgow & Leeds), 11, p. 9. (http://www.spunk.org/texts/pubs/hn/sp000025.txt).

ROGERSON, R. J. (1999) Quality of life and city competitiveness, *Urban Studies*, 36(5/6), pp. 969–985.

SAYER, C. (1992) The City of Glasgow, Scotland: an arts led revival, *Culture and Policy*, 4. (http://www.gu.edu.au/centre/cmp/4-06-sayer.html).

SPRING, I. (1990) *Phantom Village: The Myth of the New Glasgow*. Edinburgh: Polygon.

STEVENSON, D. (2003) *Cities and Urban Cultures*. Maidenhead: Open University Press.

TUROK, I. and BAILEY, N. (2004) Twin track cities? Linking prosperity and cohesion in Glasgow and Edinburgh, *Progress in Planning*, 62, pp. 135–204.

URBAN TASK FORCE (1999) *Towards an urban renaissance*. London: E & FN Spon, HMSO.

VOLKERLING, M. (2001) From cool Britannia to hot nation: 'creative industries' policies in Europe, Canada and New Zealand, *International Journal of Cultural Policy*, 7(3), pp. 437–455.

WILLIAMS, D. (1996) *Creating Social Capital: a Study of the Long-term Benefits from Community Based Arts Funding*. Adelaide: Community Arts Network of South Australia/Australia Council of the Arts.

WISHART, R. (1991) Fashioning the future: Glasgow, in: M. FISHER and U. OWEN (Eds) *Whose Cities?*, pp. 43–53. London: Penguin.

WOOD, P. (2002) *Releasing the Cultural Potential of Our Core Cities*. Stroud: Comedia.

Appendix 1. Interviewees and Participants in Focus Groups

In selecting interviewees and focus group participants, an effort was made to ensure a balance between narratives about Glasgow at the time of bidding for the ECOC and the event lead-up (1986–89), during the event implementation (1990), its immediate aftermath (1991–95) and its medium- to long-term legacy (1996–2003). Hence, the individuals interviewed have held their positions (as artists, politicians, etc.) throughout all or some of these periods.

Interviewees were drawn from

(1) *Local authorities, public agencies and quangos*: Glasgow District/City Council;

Strathclyde Regional Council; Convention of Scottish Local Authorities (COSLA); Glasgow Action; Scottish Development Agency/Glasgow Development Agency/ Scottish Enterprise Glasgow; Greater Glasgow Tourist Board.

(2) *Opinion leaders*: Freelance journalists and art editors; academics researching the ECOC programme, cultural policy, city regeneration and/or city marketing; and local activists/writers.

(3) *Creative groups*: Grassroots arts organisations; established arts venues; young artists; and event organisers.

Seven focus groups were convened from the following

(1) Journalists;
(2) community arts representations;
(3) business tourism operators;
(4) visual arts workers;
(5) venue-based institutions;
(6) creative industries; and
(7) Glasgow City Council.

Appendix 2. Methodological Approach to Press Content Analysis

Content analysis has been defined as

> an approach to the analysis of documents and texts ... that seeks to quantify content in terms of predetermined categories and in a systematic and replicable manner (Bryman, 2001, p. 177).

In a well-known definition, Berelson (1952, p. 18) adds that the technique implies "the objective ... quantitative description of the manifest content of communication". The reference to 'objectivity' implies that rules and codes are clearly outlined *a priori* for the transfer of the original material into categories. However, a limitation is the focus on counting text rather than analysing content. Without an interpretative frame, the analysis cannot go beyond "quantifying the most straightforward denotative elements in a text" (Ahuvia, 2001, p. 139) and thus cannot touch on the latent meanings and implications of the material

under review. In order to address this limitation, the CCPR project combined established quantitative techniques (centred in coding objective states such as date of publication, etc. and extremely useful to manage large datasets) with a qualitative approach, focused in the identification of themes and dispositions (May, 1997, pp. 171–175). The latter required interpretation on the part of coders and led to non-mutually-exclusive categories. Furthermore, to guide interpretation, this approach required an understanding of the social context for the items under analysis.

The identification of main themes was finalised before embarking on the analysis proper but was drawn from the material itself, after field-testing by two different researchers. A draft thematic codebook was first established, based on an extensive literature review on Glasgow 1990 and the ECOC programme, and was then tested through two independent readings of a sample of articles representing different time-periods: 1986, 1990, 1995, 2003. The final codebook is presented in Table 1. Given the non-computerised nature of clippings from the 1986–91 period, we were never to record how many articles merely mentioned each theme. Instead, respective articles were coded according to their *primary* theme, identified through assessment of heading, sub-heading, first paragraph, photograph (when applicable) and/or overall article emphasis.

Each article was also coded according to the attitude the journalist appeared to take towards the key issues identified. Attitude was recorded as simply as possible, reflecting five potential categories (see Table A1). The coding of the material was conducted with the operational guideline that the primary focus was to mark the overall tone of each clipping towards the ECOC 1990. Where difficulties arose, notice was made of the headline and conclusion of the article.

Clearly, the need to establish mutually exclusive codes prevented a more complex analysis, such as the degree of criticism or praise or variations in thematic emphasis,

Table A1. The five potential attitudinal categories

Code	Category	Description
1	Neutral	Descriptive reports with no clearly discernable attitude
2	Negative, descriptive	Reports that cover news with a clearly negative attitude
3	Negative, analytical	towards the ECOC 1990. Divided between those that cover bad news or negative opinions without the journalist/paper explicitly taking a side and those that reach a negative conclusion following analytical reflection on events
4	Positive, descriptive	As above, but in relation to positive news and debate
5	Positive, negative	about the ECOC

but this enabled a clearer and still meaningful statistical analysis. It is here that the sheer bulk of material (up to 5700 articles) becomes particularly relevant, evening out and balancing most of the coarseness of approach. With each clipping becoming a statistical component of the whole, the consistency of approach becomes more significant than the numerical reduction forced upon each individual clipping. The balance between overarching statistical analysis and the interpretation of individual press clippings is a central element of our methodology that, ultimately, allows assessing a large dataset over an extended period without artificially reducing it to explicit profile information, but rather accounting for other contextual factors such as the inclinations or attitudes of journalists and the papers they represent.

For a fuller account of the methodology developed to analyse the newspaper archive, see Reason and García (2003).

Urban Designscapes and the Production of Aesthetic Consent

Guy Julier

Introduction

Literature on place-making is, by now, copious. A body of publications emanates from architectural studies that considers the built environment in terms of urban forms and planning that differentiates and distinguishes locations. A more recent corpus is directed towards marketing, tourism and business management; this is preoccupied with the development of branding programmes to identify, articulate and mediate the 'unique selling propositions' of locations.

The central concern of this article is in the critical analysis of a wider range of design activities that take place in place-identity formation—those between 'landmark buildings' and beyond the branding programmes. It reveals the network of interests that link design production, regulation and consumption within urban locations and how place-specific design identities thus follow. In piecing these nodes together, the concept of 'urban designscapes' is presented as a term to express the network of activities and artefacts that produce place-identiy within cities. This is about the 'things' that make up their fabric as much as their representation through mediatory forms and the symbolic role of its participants. Thus the article moves beyond a-critical notions of culture-led urban regeneration to question *what kind of regeneration* is being foregrounded in specific instances. How is design mobilised to affect the urban habitus or, otherwise, both the individual and collective dispositions and practices that structure cultural capital? Thus the material objects of design come into view, but also the narratival cues that accompany these and give meaning to identity within regeneration. These may be programmatically organised through, for example, place-branding initiatives or may come about through a confluence of activities.

Three contrasting case studies are used to interrogate this process. While in each case design is mobilised within a response to post-industrial regeneration, this is done with varying degrees of programmatic, top–down direction. Barcelona is perhaps the most oft-cited example of design-led urban regeneration and yet its internal dynamics and meanings are, I believe, only very superficially understood and are invariably misrepresented as being far more programmatic than, in practice, actually happened. Nonetheless, the Barcelona example provides a rich case study of a fit between design production and consumption that was regulated through the interests of an hegemonic, avant-garde élite. It therefore provides a useful conceptual model to compare with the two more focused and shorter case studies of Manchester and Hull. Manchester is chosen as it demonstrates a much more self-conscious attempt to orchestrate place-identity through the appropriation, channelling and regulation of a specific aesthetic outlook for the city, drawing on pre-existing designerly and attitudinal resources. The case study of Hull is even more self-conscious and programmatic in its place-identity creation; it also draws on pre-existing circumstances but this attempts more overtly to *produce* an aesthetic aspiration for its own citizens. For Barcelona and Manchester, the role of a designerly élite is foregrounded in understanding the development of place-identities, both in terms of their influence and the symbolic capital they embody for those locations. The exposition

of Barcelona's designscape is considerably lengthier than that of Manchester and Hull. This is testimony to its density and extent of its design infrastructure, while by contrast, Hull's is largely nascent.

The research included close (and pleasurable) observation—or as design ethnographer Judy Tso (1999) terms, 'deep hanging out'— of the day-to-day activities, locations and artefacts of design production and consumption in the three locations. An archive of some 40 taped interviews I undertook with designers, manufacturers, retailers and design officials and commentators in Barcelona, 1988–92 revealed the informal networks between them and their interfaces with institutions and mediatory forms of the city's designscapes. Interviews, press releases and brand reports provided data in Manchester and Hull. Ultimately, though, my chief interest is in developing a critical perspective rather than in an exhaustive empirical account.

This article begins by challenging architectural criticism that relies on form alone to discuss place-identities. It is equally necessary to problematise the contributions of marketing and branding to this process. This in turn leads us towards the 'urban designscape' notion as a useful tool for analysing the dynamics of design within culture-led urban regeneration before its application to the three case studies.

Between Buildings

Within architectural studies, Kenneth Frampton's promotion of 'critical regionalism' in the early 1980s revived the issue of the relationship of built form to regional identities as an anti-centrist and anti-modernist/International Style conception (Frampton, 1986). In practice, while Frampton's position may have sparked healthy discussions around the issues of place-identity and architecture, the aesthetic outcomes continue to be sporadic and isolated. For example, the activities of the circle of Imre Makovecz and his circle or 'Hungarian Organicists' provided an engaging example of a self-conscious attempt to revive architecture within a place-making process. Makovecz's interest was in an architecture that drew on the skills of local craftspeople and expressed the historical traditions of a location. But it is noteworthy that this movement's approach only came to international attention via their Hungarian pavilion for the Seville Expo of 1992. Indeed, consonant with the 150-year tradition of world expos (stretching from the 1851 Great Exhibition to the 2000 Hanover Expo), their national pavilions have perhaps been the only platform where place-identity is so clearly marked through the architectural signifier.

Beyond such occasions, the tendency—certainly in terms of 'landmark' new buildings—has been towards a globalised civic patronage by major urban centres commissioning global 'name' architects (such as Frank O. Gehry in Bilbao, Prague or Los Angeles, Daniel Libeskind in Salford, New York, Berlin, etc.). Thus a location is distinguished as much by its civic patronage (having the money and taste to engage a particular famous architect) as by the building itself. Needless to say, this consideration takes us rapidly to Bourdieu's notion of cultural capital. It is a location's disposition to employ a certain architect, for example, that is important. The buildings symbolise that disposition, that sensibility, that attitude. They are 'objectivated cultural capital' (Leach, 2002, p. 283).

In thinking about the role of architecture in the identification with space, architectural theorist Neil Leach takes us beyond the object to, "engage the subjective processes of identification" (Leach, 2002, p. 281). We can begin this analytical approach, Leach argues, by considering Bhaba's concept of nation as 'narration' (Bhaba, 1990). Here, the meaning of nation comes into life through the language and rhetoric that articulate it. This narration is not abstractly independent of objects, however. Instead, what is said is activated by objects, while narration also contextualises and gives meaning to them. Design objects and discourse are mutually dependent in generating belonging. In addition to this concept of 'narrating' place-identity, Leach also draws attention to how this is 'performed' by drawing on Butler (1990). Within this conception, identity is subjective. It can be played

out through the reiteration of a set of norms. These norms, then, can be adopted or appropriated, used, used up and relinquished.

In respect of the built environment, Leach provides a useful starting-point for thinking about how identity is produced, both internally for a community or population and for their external audience. His foregrounding of architecture as a dominant aesthetic focus within this process raises, however, an important question for this article. We are left wondering, first, what role other aesthetic forms such as urban furniture, city council or regional government websites, tourist literature or the retail and leisure infrastructure, play in this process? Stemming from this, we might ask what are the mechanisms by which narrations and norms for the playing-out (or, otherwise, performativity) of place-identity are generated, circulated and adopted? Is there a link between the two questions? In other words, how does design, beyond architecture, contribute to providing official or non-official 'stories' and 'ways of doing things' that, in turn, fashion specific aesthetic outlooks for a place?

Beyond Branding

Branding has become a central motif of contemporary design practice in recent years. Its application, largely through tourism marketing, to the definition and communication of the characteristics of locations suggests an alternative line of enquiry beyond architectural criticism. And yet, as a process it is beset with operational problems that require its analysts to regard it in concert with other design manifestations to understand its workings.

Place-branding is the process of applying the branding process—as applied to commercial products—to geographical locations and is a burgeoning activity within advertising and marketing (Olins, 1999). The relationship of product–country image is claimed as the 'most-researched' issue in international buyer behaviour; there are, it is calculated, over 750 major publications by more than 780 authors who address this theme

(Papadopoulos and Heslop, 2002, p. 294). Specialist place-identity marketing and brand consultancies have emerged (such as Total Destination Management in the US and Placebrands Ltd in the UK). 'How to do place branding' books have been published (for example, Morgan *et al.*, 2002; Kotler *et al.*, 2002; Ashworth and Goodall, 1990).

Several authors recognise the inherent problems in directly applying the notion of product branding to places (for example, Anholt, 2002; Papadopoulos and Heslop, 2002; Bennett and Savani, 2003). First, a place is not a primary, singular product, but an agglomeration of identities and activities. While identifying, articulating and nurturing these, they nonetheless often add up to the most generic of brand values in terms of place-wide marketing. Internally to a location—for its population, that is—these values may not necessarily reflect or promote their reality. Externally, to the rest of the world, they appear bland and undifferentiating. A web-survey of global locations' city authorities bears out the homogeneity of brand-value language that results. In 2004 Singapore, Brisbane and Birmingham all described themselves as 'dynamic' and 'cosmopolitan' or 'diverse'. Johannesburg and Manchester both came in with 'vibrant'. Birmingham, Glasgow and Johannesburg were 'cultural'; Santo Domingo and Brisbane, they claim, were both 'sophisticated'.

Secondly, the process is not simply one of 'rebranding' but more of 'brand management'. It is about the "slow moving husbandry of existing perceptions" (Anholt, 2002, p. 232). The branding of a location may be read as an attempt to create and nurture the narratives that give meaning to a place. Extending from Leach's approach to the interpretation of architecture, place-branding self-consciously provides linguistic cues to outsiders and citizens through and from which the material attributes of a place are perceived. Thus—for example, a brand-value of a place as 'vibrant and cosmopolitan' immediately appears to be implicit in the buildings or restaurants that are encountered. The job of the brand manager is to find ways

of orchestrating these pieces of pre-existing aesthetic information.

This nurturing of urban identities is not necessarily as programmatic as brand consultants would perhaps like. While clear, easy-to-understand narratives of a place may be easy to develop and communicate in themselves, they also require confirmation in the material circumstances to which they relate. This creates an uneasy relationship. It requires the participation and aquiescence of a range of stakeholders—the brand values have to be understood and reiterated by citizens. Policy implementation in the husbandry of a place-brand may be constrained so that it can only be undertaken in limited areas. Changes of local, regional and and national governments and their policies may impact on developments. Key personnel in the process may leave their posts, interrupting the brand management processes. Economic fluctuations and financial constraints may hinder developments—places may not be so 'vibrant' in a recession. Unforeseen circumstances in relation to the cultural or material fortunes of a place may sabotage the process: city football teams may have disastrous seasons (for example, Leeds in 2003–04); opera houses sometimes burn down (Barcelona, 1994; Venice, 1996).

In this context of uncertainty, 'hard-branding' the city is a precarious strategy. Evans (2003) uses this term, 'hard-branding', with reference to the impact of the creation of large cultural schemes—*grands projets* such as new museums, arts complexes, theatres or opera houses—on a wider strategy of regeneration and place identification. He sees this strategy as imposing clear signifiers of modernity into the post-modern cityscape. As with the above discussions in marketing, there is a similar alertness to the problems of these enterprises: the danger of a 'me too' copycat reproduction of signature architect-led projects across a range of global cities ultimately homogenises identities of global cities; the appropriation of public practice into a realm of private consumption effectively conspires to commodify culture further; the accompanying gentrification processes of cultural quarters marginalises local residents; the threat of brand decay as attractions lose their shine makes them short-term investments; and the potentially crippling costs of such ventures makes them far from risk-free.

Evans does not fully explain his use of the term 'hard-branding'. However, the implication (by reference to Mommaas, 2002, p. 34) is that the development of contemporary cultural infrastructure is an inflexible and imposing strategy in which high coherence and order are sought through the creation of clear, readable messages which in turn make, for an audience, choosing an easier process. This usage does indeed have some resonance with its meaning within the design and marketing industries. Conversely, the term 'soft-branding' is used to denote a looser system in which a broader palette of options is available to carriers of brand identity (for example, M.R., 1998). This is sometimes found in tourism and leisure industries where, for example, a hotel or a campsite pertains to a parent brand, mostly for the sake of marketing, but retains its individual identity beyond this. The parent brand therefore denotes a particular level of service content and quality, but does not dictate to the operational systems of its parts as to how these are achieved or as to the more nuanced aspects of their aesthetic dimensions.

The notions of hard- and soft-branding allow a useful tool to critique and understand the cultural role of design in urban regeneration. Hard-branding may come into play when inventing a place a-new. The Disney-created community, Celebration in Florida, might be an example of this. It may work in the context of a thoroughly compliant population and bottomless budgetry resources, but its limitations are not difficult to spot. This distinction is, in fact, more useful when thinking about audience and context. At a basic level, the hard-brand may be used to distinguish a location at 'entry point'. So 'Gaudì's Barcelona' becomes an associational trigger of expectation of what the city is about. Gaudì's Sagrada Familia may be an immediate shorthand to identify Barcelona, perhaps

as 'avant-garde' and 'creative'. However, beyond this are many other layers of activity that overlap with this, but also say other things too.

In considering the emergence or positioning of place-identities via the mobilisation of the symbolic capital of design, we must therefore not restrict the analysis to visual identity programmes and *grands projets*. Instead, the full range of design production and consumption has to be considered. In this way, the coherences and contradictions—and, indeed, the contradictions that form part of those coherences—within their practices and discourses add up to an extended field of consideration.

An early step in this direction was taken by Molotch (1996) in his essay on 'L.A. as design product'. Molotch recognises the difficulties in identifying a coherent design aesthetic within a freewheeling city such as Los Angeles. However, his quest was to show how local aesthetics are important to the business climate or 'industrial atmosphere'. In this, he ranged through the furniture design of Harry Bertoia and Charles Eames, the automobile design to come out of southern California studios of Japanese companies (such as the Isuzu Trooper or Toyota's Lexus), the effects of the movie industry, tourism and fast-food. His conclusions were that the LA style is not characterised necessarily by a particular 'look', but by a sheer volume of cultural producers and, perhaps, some shared sensibilities about being renegade but also versatile. Molotch's approach is far from being empirically tested. There is no real evidence that the examples he cites do add up to a self-identity. His account is production-led, with no reference to LA design's impact on the consumer culture of the location. While he mentions the importance of art colleges as a possible bonding agent within this system, there is little evidence of how or if any of these disparate design activities add up in any way. Neither is there any evidence of this being a reflexive process, either within the design industry itself (as in 'the deliberate designing of California-ness into artefacts') or through any place marketing initiatives. Nonetheless, Molotch is hinting at the notion that an urban habitus is performed through a series of aesthetic platforms.

More recently, Bell and Jayne (2003) mobilise this approach in a more instrumental than analytical way. Drawing heavily on my own concept of a 'culture of design' (Julier, 2000), they map out their aspirations for north Staffordshire in terms of design-led regeneration. Thus they suggest that a nurtured co-existence of producers, consumers and designers in a coherent circuit of culture, can in turn generate economic, social and cultural value for a location. Their scope in terms of design ranges across crafts, product engineering, retail and entertainment, public spaces and architecture with the city as the cradle of resultant entrepreneurial networks and consumer practices feeding off each other. The interlocking of such processes within a densely operating, localised framework would create what I term an 'urban designscape'.

Barcelona is oft-cited as a beacon of regeneration and identity-formation through urban design (Hajer, 1993; Urban Task Force, 2000; Bell and Jayne, 2003; McNeill, 2003; Marshall, 2004). Throughout these studies, the regeneration of public spaces and services is analysed almost entirely as a functional activity concerned with delivering more efficient, cleaner and attractive environments. They do not connect these developments to a wider change in the taste patterns of a population, or at least, a change of aspiration. In short, public consumption is not connected to the private and vice versa. There is an assumption that the aesthetic element of these is neutral and that this does not play a distinguishing and defining role in the regeneration process (Crilley, 1993; Hubbard, 1996). A consideration of the term, 'urban designscapes' provides a conceptual model for looking at how public and private consumption are connected within the framework of design-led regeneration, how the actor networks of agglomerations produce aesthetic consent and what kind of aesthetic consent this might be.

Urban Designscapes

The term 'urban designscapes' is intended to convey the pervasive and multilevel use of the symbolic capital of design in identifying and differentiating urban agglomerations. Hence it exists through a variety of aesthetic platforms, ranging through brand design, architecture, urban planning, events and exhibitions. But it also extends to the productive processes of design policy-making and implementation, design promotion and organisation and the systems of provision of design goods and environments within these contexts. The actor networks that make up the design culture of a location therefore take on a symbolic role—they become a kind of 'meta-activity' that frames and explains urban social, environmental and economic identity. However, this identity-forming process also purports to extend into the consumption sphere of the city. Lifestyles, taste patterns and everyday practice supposedly become attitudinal markers of an urban habitus.

This notion of urban designscapes is adapted in the first instance from Chatterton and Holland (2003) who employ the term 'urban nightscapes' in their book of the same name. They analyse the interdependencies of youth cultural activities, urban night-time economies and the corporatisation of the public sphere. Thus the 'fit' of productive and consumer interests in image-building is brought into view. Within marketing studies, we have seen the recurrent use of the term 'servicescapes' in terms of this notion of 'fit'. Originated by Bitner (1992), this has largely come to be understood as incorporating consideration of the built environment of a service delivery point but also the 'atmospherics' of the location. Thus ambient details such as furnishings, noise, music and air quality are factored into the appreciation of what facilitates effective service delivery. The judgement of effectivity of these is not just from the point of view of the ability of users to fulfil their tasks within an environment. It also means the delivery of a coherent and understandable message about itself and its differentiation from other competitors through that environment. In other words, the design of a 'servicescape' engages both functional and symbolic demands.

This 'servicescape' concept may be common-sensical, especially to the seasoned retail interior designer whose brief is to orchestrate not just spatial form but lighting, soft furnishings, signage, background music or employee uniforms within the bounds of a brand identity. But the emergence of the term in the early 1990s signals several important shifts for how design has come to be played out. In the first instance, the interest of such academics in disciplines of retail and marketing reflects a growing more general design consciousness and ubiquity of reference to it. Secondly, it represents a shift in acknowledgement of design in the 1990s from being about the shaping of individual products, spaces and images to the wholesale aesthetic orchestration of systems. By extension to this argument, Lash notes that

> Culture is now three-dimensional, as much tactile as visual or textual, all around us and inhabited, lived in rather than encountered in a separate realm as a representation (Lash, 2002, p. 149).

He goes on to argue that we have become an architectonic, spatially based society and information is reworked in these planes. Culture is no longer one of pure representation or narrative where visual culture conveys messages. Instead, I would add, design culture formulates, formats, channels, circulates, contains and retrieves information through a number of channels. Design is more than just the creation of artefacts. It conspires in the structuring of systems of encounter within the visual and material world. Collectively, these add up to the creation of urban designscapes.

It is important, though, not to get carried away into vague assertions on the state of aesthetics and the everyday lifeworlds of cities. The work of such sociologists as Lash and Urry (1994) and Featherstone (1991) is important in recognising the growth of cultural goods and the increased importance of

the global sign economy. Scase and Davis (2000, p. 23) take this notion further to claim that the creative economy is at the "leading edge of the movement towards the information age [as] their outputs are performances, expressive work, ideas and symbols rather than consumer goods or services". They are paradigmatic of broader changes in economic life. These writers, however, provide little empirical evidence to demonstrate this or illustrate the mechanisms by which it comes about (Nixon, 2003). Equally, the term 'urban designscape' may signal an automonous state of being, somehow hovering over the city. It is important, therefore, to delve deeper and analyse how this designscape functions in terms of the interests that work within it and the dispositions that are foregrounded by it.

Any form of '-scape' involves a privileged position in terms of what it defines for viewing. Thus, drawing on Zukin (1991, 1995), we may regard a 'landscape' as requiring a hegemonic position that reflects economic, cultural and social priorities of power. A landscape has to be seen from a vantage-point. Zukin's chief interest was in the transformation of urban spaces and the absorption of either vernacular or avant-garde practices into a dominant value system. Within the term 'designscapes', I infer an extended system that engages not just spatial attributes, but also issues of taste, practice and the circulation of design that are nonetheless still inflected by power mechanisms.

Revisiting the 'Barcelona Paradigm'

The 'Barcelona effect' of the 1980s and 1990s came about through the interraction of several layers of design activity and consumption. These were all connected by both formal and informal networks that coursed between governmental policy-making, design promotional institutions, associations, civil society and the industrial and retail infrastructure for design. Furthermore, a range of mediative systems for design products and discourses existed, from local press and TV coverage, to design festivals, to emergent new bars and restaurants.

Thus while a certain designerly élite held some hegemonic power in the shaping of a new urban habitus, there was also a constant ripple effect outwards from it. Examples of how these networks functioned across Barcelona are too numerous to describe fully here, but a few may give a flavour of how they acted, the coincidences of personalities and their impact.

For the historian of Barcelonese design, the autobiographies by two key figures of its architectural and design milieu provide a stunning 'who's who' of its actor networks. Oriol Bohigas' *Combat d'Incerteses* (1989) maps his pre-eminent position within the development of a Barcelonese sensibility towards urban planning and regeneration. During the 1980s while the renovation of public space was orchestrated according to a city-wide plan, it was characterised by small, particularist interventions. This stemmed in part from an anti-modernist, totalising discursive position that was shared across design disciplines in Barcelona (Julier, 1990). But it was also the result of a considerable amount of interchange between the City Council's planning department—with Bohigas in charge 1980–84—and civil interest-groups such as neighbourhood associations (Calavita and Ferrer, 2000). It was also the result of long-standing, close relationships Bohigas had with other architects and politicians that facilitated swift implementation of his plans. *Combat d'Incerteses* thus tells the story of a dense network of architects, designers and politicians and their friendships, debates, informal and formal points of contact.

The second key figure whose autobiography we could plunder is André Ricard and his *En Resumen* (2003). Ricard occupies an historical position in the institutional and professional development of design. He more or less single-handedly introduced the concept of industrial design to Barcelona in the late 1950s. Through the 1960s and 1970s, he was at the hub of design promotion and the foundation of Barcelona's Design Centre, BCD, in 1976. He designed Barcelona's successful canditature presentation for the 1992 Olympics and its Olympic torch. He was also

chair of a steering committee that wrote a design policy for the autonomous Catalan regional government in 1984 (Generalitat de Catalunya, 1984). Thus, like Bohigas' autobiography, his book comes alive in its onomastic index. Here, we see the evidence of the networks of designers, politicians and industrialists that were engaged in his project to champion design in the lists of names and representatives that are familiar to either the historian of Catalan design or those already 'in the know'. In both Bohigas' and Ricard's accounts, we see the mechanisms of a cultural élite at work in a highly localised and dense framework of activity.

This largely upper-middle-class élite also extended its network outwards. For example, the creation of the Barcelona Centre of Design—a design centre to promote the profession within industry—is noteworthy for its implication into Barcelonese commercial and civil society. Its council included representatives of institutions ranging from neighbourhood associations, heritage lobbyists, savings banks, the press association as well as the city council and regional governments. Such connections took the institutional organisation of design promotion beyond its own élite.

At the level of consumer culture, Barcelonese designers were indeed closely connected into the fabric of the city. The more prominent emergence of design retailers in the 1980s such as Vinçon, Pilma, B. D. Ediciones de Diseño and Santa & Cole ensured the high-street visibility of Catalan design production. All four of these had their distinct approaches to retail. However, what binds them is the fact that the vast majority of their goods, in particular furniture, were dominated by Barcelonese design products. B. D. Ediciones de Diseño and Santa & Cole both produced their own commissioned new designs and also re-edited historical 'classics' by named designers such as Antoni Gaudì and Charles Rennie Mackintosh. In either case, these, along with Pilma and to a lesser extent Vinçon, provided important taste nodes of high design for citizens. As a female Barcelonese primary school teacher once told me

> I go to Vinçon to see what there is, what's *in*, and then I go to the [the department store] El Corte Inglés to buy something a bit like what I see there but more affordable.

However, even if citizens did not visit these retailers, their effect spilled out into the streets in any case. B. D. Ediciones de Diseño provided many of the city's public benches (Oscar Tusquet's 'Banca Catalana') and Santa & Cole produced and sold much of its park lighting (Beth Galì and Marius Quintana's 'Lámpara Alta').

Design consumption could also take place at a spectacular and yet fleeting level. The plethora of new bars during the late 1980s all featured furniture, lighting and interior design by 'name' Barcelonese designers. The connections were taken a step further as several were owned or promoted by design figures. Best known of these was the Velvet bar which opened in 1986. It was the creation of Juli Capella and Quim Larrea who also founded and edited Spain's premier design magazine *Ardi* in 1988. Capella and Larrea went on to establish Barcelona's bi-annual design festival, the Primavera del Disseny in 1991. They also curated the 'Barcelona House' for the 1992 Olympic Games cultural programme—a house furnished with bespoke objects by Catalan designers that were subsequently put into production.

The above slice through the Barcelona designscape illustrates the density of actor networks and the close relationship between individual *animateurs* of design and the institutions of civil society but also how these were materialised into the urban fabric. This density was highly visually evident within the city but also insitutionalised into the everyday relationships of practitioners within the design milieu. As one Madrid-based designer told me in 1990

> I have had plenty of opportunities to set-up my practice in Barcelona, but, really, I'd find it too claustrophobic. In Barcelona, if you haven't been in touch with other designers for a couple of weeks you have to make up some excuse like your wife has been ill or you've been away on holiday! (P. Miralles, industrial designer; personal interview 25 September 1990).

Thus, this milieu displayed those features of 'network sociality' redolent of the creative industries that McRobbie (1998, 2002) identifies. In Barcelona, this was to some degree about a closed institution in which its actors were constantly required to maintain their social capital. The creative workers that McRobbie discusses are directed towards maintaining their value in a globalised marketplace. The milieu under discussion here, though, was directed towards Barcelona first both in terms of their marketplace and their social horizons, and Catalonia/Europe secondly.

Figure 1 shows the networks that made up Barcelona's designscape, taking into account key individuals, locations and organisations. It illustrates how the interests of individuals and organisations were materialised and mediated, providing specific points of encounter for a wider public. It is noteworthy, following the diagram, that few of the mediatory forms and sites of consumption for such encounters had direct relationships with the institutions of governance and civil society. This may seem an obvious point; after all, one would expect the designers themselves to develop their formal content. But the diagram also locates the role key designers and design institutions as interlocutors between these points. Meanwhile, it reveals that the most public of sites of encounter, urban design and design exhibitions, provided interfaces between both designers and local government. These were key moments in the shaping and mediation of aesthetic content. What were the nuances and meanings of this content and how it was converted into specific narrations of place require further consideration.

While the designerly aspirations of these professionals became more pervasive in Barcelona through the 1980s, these were also fixed at a refined level (Narotzky, 2000).

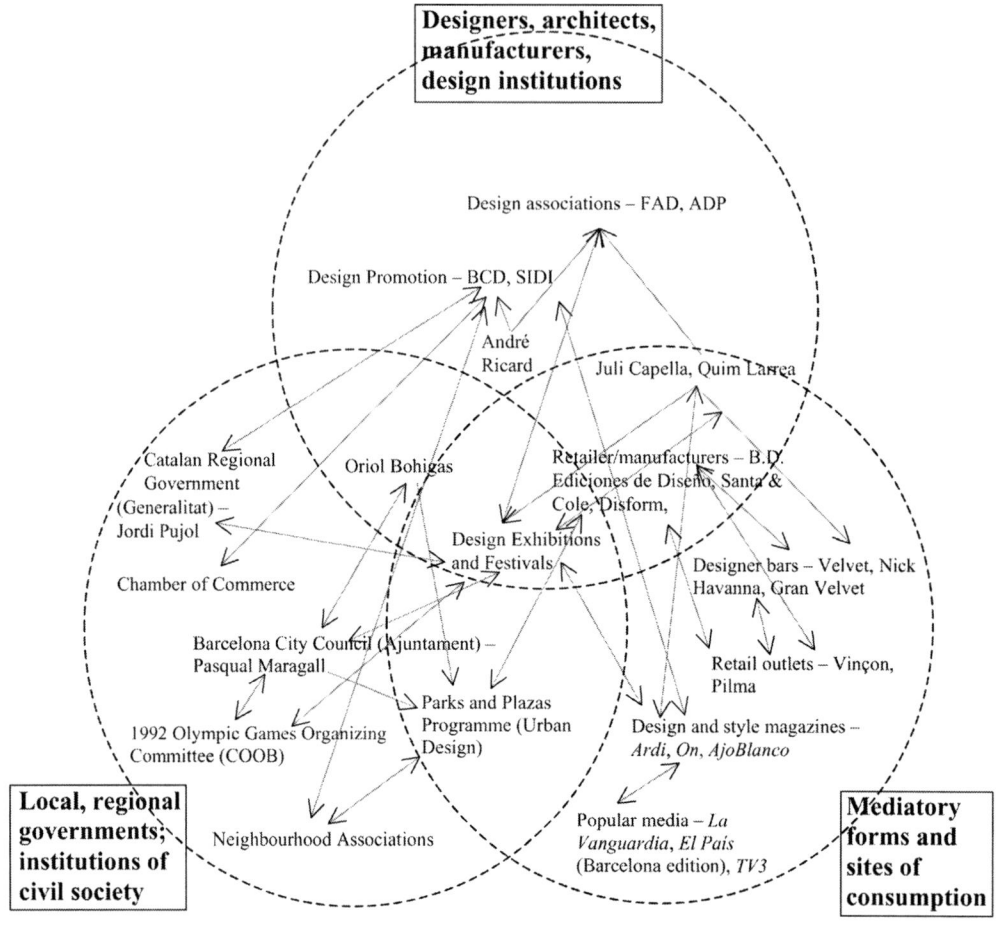

Figure 1. Domains of design culture that formed Barcelona's designscape in the 1980s and 1990s. Arrows denote lines of *direct* relationship and influence.

There was little watering-down of its content for a wider, popular audience. Modern Catalan design emerged from the late 1950s with a clear focus on exclusive, modern design for domestic consumption—in particular for the furnishing of trendy Costa Brava villas or Barcelona penthouses (Julier, 1990). In the 1980s, a broader marketplace was found for it, but the design artefacts and the milieu that produced them maintained an air of exclusivity. The increasing ubiquity of design in 1980s Barcelona reflected certain ambitions of the Barcelonese upper middle class. Here, design

expressed the taste and aspiration of [this] social élite, based on an idea of modernity distilled from the creative energy of turn of the century *modernisme* ... [that displayed a] quintessential spirit of modernity, democracy, progress and avant-garde culture (Narotzky, 2000, p. 241).

Needless to say, the emergence of this new designscape coincided with the revindication of Catalan bourgeois identity within the context of the post-Francoist transition. As already highlighted, a designscape implies a dominant construct of habitus. But it engages a criss-crossing of activities that are impossible to co-ordinate in any programmatic way. Instead, they are all actioned within a binding, but not necessarily overtly articulated, ideological and aesthetic framework.

From this élite, this particular designscape rippled outwards into the individual and collective consumption practices of Barcelona's citizens. Consumption studies have, to date, focused primarily on its private, domestic and individual modes (see, for example, Campbell, 1987). This is particularly so within design history and design studies (Julier and Narotzky, 1998). Where perhaps this individual expression has been linked into collectivities has been in terms of the consideration of communities around particular leisure pursuits (Hebdige, 1988), lifestyles or brands (Muñiz and O'Guinn, 2001). Narotzky (2000) skilfully demonstrates, though, that an effective interaction between private and public systems of provision existed in the Barcelonese context. She draws on Kopytoff (1986) who wrote of the 'singularisation' of non-commodified collective goods that are placed beyond the sphere of exchange. This is achieved, for example, through their taste-brokers such as historical commissions, neighbourhood associations or panels deciding on public monuments. But this could also take place in the frequent exhibitions of Barcelona design or even in the display of individual pieces in design retail spaces. In Barcelona, this process was also manifested in the public sphere of urban design and designer bars where the tastes of this élite were placed as what Narotzky calls 'secondary commodities'. These spaces provided a backdrop to consuming other things, "such as food, drinks, leisure services or public spaces" (Narotzky, 2000, p. 234). Thus new bars would feature Catalan design objects—the higher cover price on drinks would subsidise this ephemeral design consumption. So whilst this new design originated in the framework of high culture and bourgeois taste for the modern domestic interior, in the 1980s it expanded its sphere of influence into the public sphere. New public spaces and leisure services became pervasive and sometimes (but not always) persuasive loci for the incalculation of a new urban habitus.

Such events and the personalities attached to them were unavoidable in the city. If missed on the street or the bars, they were accessible via front-page reporting in Barcelona's media outlets of *La Vanguardia, Avui* or *TV3*, or more consistently the newly established or re-established design and style magazines such as *On* (1979), *DeDiseño* (1984), *Ardi* (1988) and *Ajoblanco* (1988). Domestic taste and modernity were also linked to the changes in the fabric of the city via a soap opera in 1992. 'Poblenou' featured a family who, having won the lottery, moved from their run-down home in the Gothic Quarter to a brand-new designer apartment in the Vila Olímpica (Campi, 2003).

Of course, this designer version of Barcelona did not go uncontested. For example, the appearance of the highly minimalist Plaça dels Països Catalans in 1984 provoked considerable, vehement objections from local neighbourhood associations for its cold, hard and overmodern aspect. Calavita and Ferrer (2000) trace the political origins to the renovation of public spaces during the early 1980s in the pressure exerted by urban social movements of neighbourhood associations in the city dating back to the early 1970s. Indeed, Bohigas himself and many of his circle were closely involved in these. The neighbourhood association objections to the design of this plaza may appear, then, an ironic twist. Moreover, this underlines the contingency of this élite on popular engagement with their project, whether it be critical or accepting.

By revisiting the emergence of a design-led urban identity that existed within a discourse of regeneration, we can see in the first instance how it emanated from a specific urban élite. This upper-middle-class élite promoted its take on modernity and avant-gardism through design practices that confirmed prevailing ideological currents. These involved the grooming of a specific form of Catalan modernity oriented towards its place in Europe. This project was characterised by its 'network sociality' that extended beyond the operational interactions of designers in their everyday professional lives into local and governmental policy-making and to the institutional representatives of civil society. Equally, then, the aesthetic tastes that materialised the élite's ambitions were carried from the private sphere of design for domestic consumption to the public sphere of design for civic engagement, from individual to collective consumption. Their mediatisation through exhibitions, news reporting and style magazines but also proclamations of support made by politicians helped to develop a narrative to contextualise them. In turn, this process produced an urban designscape that, at least from the outside, appeared to have remarkable coherence. The activities added up to a design-led urban identity.

Given this density of design activity, Bell and Jayne (2003, p. 127) admit that it is not possible to say that there was a programmatic ambition to promote design or even a particular form of design on the part of local authority policy and planning in Barcelona. Those engaged at the nub of design promotion regarded politicians' interest in their activities largely as opportunism (A. Ricard, personal interview 7 June 1988; M. Felip, Director of the Barcelona Centre de Disseny, personal interview 7 July 1988). Politicians in themselves were at least keen to be seen to be in the design-fray. However, the range and extent of design activities was too varied and widespread for any centralist organisation, as they belonged to individuals, small companies and professional associations as well as local government departments.

Barcelona has never, in fact, been subject to any place-branding exercise internally or by any external design consultants on the scale that is prevalent today in other locations (and as we shall see in the cases of Manchester and Hull). Indeed, much of the development of the Barcelona paradigm took place before the notion of branding places took hold in the later 1990s. If any specific 'Barcelona brand' is discernible, then this evolved out of the designscape through a shared understanding. It was these various activities that produced a 'soft-brand' or, otherwise, 'aesthetic consent'. A popularisation of the design élite's sensibility into a broader aesthetic may have been translated into a more reflexive, self-conscious value of 'Mediterraneanism'. This is evident in the *ad hoc*

campaign originated in 1984 by Pepe García, 'Barcelona Més Que Mai' (Barcelona more than ever). Here, a Miró-esque pastiche of painterly primary colours on a white background signals both a local artistic heritage and a brightness and clarity. This kind of motif was repeated in Josep Maria Trias' 1992 Barcelona Olympic logo. Equally, signifiers of Mediterraneanism were laced into the urban design of the Vila Olímpica. For example, the main street axis, the Ronda, included columns set between palm trees, suggesting the city's Mediterranean connection with ancient classicism (Campi, 2003).

The 'Barcelona paradigm' is highly specific to an historical moment that mustered both internal and external cultural, political, social and economic forces in the production of design-led regeneration and urban identity. To understand it fully, one must delve into the micro-networks of everyday relations between designers and others, as well as between design production systems and the reception of design. By doing so, we can gather a nuanced understanding of the politics and practices of design and their relationship to urban identities. We can see how expectations of the new Barcelona were articulated and disseminated and how regeneration was performed by élites and citizens alike. The design artefacts that signified urban generation were set in their place, but these also required a discourse of modernity, avant-gardism, democracy and regeneration to contextualise and filter them. In reviewing the development of urban identity in Manchester in more recent years, we can use the features that were prevalent within the Barcelona paradigm as a lever to understand the actions and uses of its own particular designscape.

Designing 'Manchesterness'

In March 2004, Peter Saville was appointed as the 'Creative Director' of Manchester. He was selected, according to the Manchester City Council's head of the Marketing Co-ordination Unit, for his, "capacity to envision Manchester as an internationally competitive city" (S. Hunt, personal interview 14 July

2004). While his brief as Creative Director was fairly open, it begins with, "images, words, typefaces, colour and all forms of presentation and he will advise Manchester on all aspects of creativity" (Manchester City Council, 2004). It would also extend to reviewing the city's urban environment and architecture. Acting as a part-time consultant, this was the first time a British municipality had appointed a senior creative with a broad-ranging and long-term remit (*Design Week*, 2004). As such, his job was to give material form to a brand strategy for the city that had been developed over a three-year period.

Saville's appointment was the outcome of a process undertaken by Manchester City Council's Marketing Co-ordination Unit that began with a review of the city's assets stemming from and as a follow-up to the Manchester 2002 Commonwealth Games. A Manchester brand identity had been created in 1998. This was commissioned by Marketing Manchester, a private–public partnership established in 1996 to promote the city and create partnerships in securing projects, events and investment. Hemisphere Marketing and Design undertook the creation of this brand identity which was synthesised into a 'Manchester Primer'. While the primer admits that, "noone wants a Draconian 'style guide'", it nonetheless provided, "joined up thinking" in terms of, "flexible toolkit elements" that give "a sense of mood . . . [to embody] . . . the Mancunian spirit". This included a 'Manchester font' that was intended to reflect Manchester's unrivalled skill at merging the old and the new, from Stockport's towering railway viaduct to the new, organic form of Urbis (Marketing Manchester, 2002).

While the Manchester Primer displays all the cues of hipness contemporary to its design historical moment—retro-1970s neo-modernist forms, photographic stock images of chilled citizens, chatty copywriting—it was loudly contested by a group of Manchester-related figures within the creative industries who called themselves The McEnroe Group (as in, "You cannot be serious!"). They dismissed the efforts of Manchester Marketing as, "dull, mediocre and worthy of a cycling proficiency badge" and challenged the City Council to improve on this (Shaughnessy, 2004). As a result, a fresh brand strategy was commissioned from Hemisphere Design and Marketing and a Creative Panel was formed by the City Council's Marketing Co-ordination Unit to advise on its further development and implementation.

This Creative Panel was partly made up of figures from The McEnroe Group. Its chair was Tom Bloxham, Group Chairman of Urban Splash. This company was a key to urban property developments in city-centre locations with a high design profile in Manchester and nationally. Other members were Scott Burnham, director of Urbis (Manchester's visitor attraction dedicated to the exploration of the city) and Rachel Haugh, director of Ian Simpson Architects who designed this Mancunian visitor attraction. Indeed, these and the other 12 members make up a veritable 'who's who' of the city's élite in architecture and design.

The shortlist that the Creative Panel put together for a city Creative Director represented a 'broad palette' of approaches (S. Hunt, personal interview 14 July 2004). It included Michael Wolff, a long-serving London-based brand strategist; Alan Fletcher, co-founder of Britain's long-standing and best-known interdisciplinary design consultancy Pentagram; Will Alsop, the renegade architect known for his masterplans for Bradford and Barnsley; and Javier Mariscal, the Barcelonese creator of the 1992 Olympic mascot. If this shortlist was to encompass a broad range of approaches to design, then its singular distinction is that all of its members were recognised and established 'name' designers. They connected the Manchester brand development into an international milieu of creatives, boosting the process's *caché*. This resonates with what Florida (2002, p. 200) calls, "the co-opting of Bohemia" in the way that renegade figures are used to provide another marketing tool in the selling of Manchester. As a surface observation, this may well hold. But a more interesting question arises in considering what *kind* of renegade fits this specific process.

To the design *cognoscenti*, the choice of Peter Saville as Creative Director may seem unusual. He is noted for his renegade approach. He is famous for his inability to make deadlines (Shaughnessy, 2004), is often highly critical of corporately or commercially driven approaches to design and noteworthy for his dedication to the personal interest of a project rather than its potential commercial gain (Poynor, 2003). Indeed, his biography reads almost like the archetype of a brilliant but unemployable designer.

But to the design *cognoscenti* in the context of Manchester's creative milieu, the choice of Peter Saville is utterly understandable. Saville was born in 1955 in the affluent Manchester suburb of Hale. He studied graphic design at Manchester Polytechnic (1975–78) where he began to design posters for the recording-label Factory, conceived by Tony Wilson (who 25 years later was a member of The McEnroe Group). Saville went on to design some of Factory Records' seminal record sleeves, including Orchestral Manœuvres In The Dark's 'Architecture and Morality' album (1981) and Joy Division's 'Blue Monday' single (1983). Indeed, the ever-sartorially dressed Saville was portrayed in the 2002 biopic feature film of Tony Wilson and Factory Records *24 Hour Party People.* Thus, Saville is implicated into the mythology of Manchester's most-known popular cultural history. During the following 20 years, Saville was noted for his rollercoaster yet flamboyant career as an art director, mostly in the fashion and music industries. His 'return' to Manchester was marked in March 2004 by a retrospective exhibition of his work—previously seen at London's Design Museum—at the Urbis centre.

A 'reading' of Saville's appointment would therefore take these biographical credentials to make him the embodiment of the Manchester brand. Of the brand values established by Hemisphere in its strategy for Manchester, the first was 'respect' that encompassed the features of 'attitude/edge/enterprise'. This is evidenced in their brand strategy report by

reference to the city's music history and, "its desire to not wear what everyone else is wearing, but to wear what everbody else is going to be wearing in six months time" (Hemisphere, 2003, p. 9). Within this narrative, Saville, or any style-conscious cultural intermediary for that matter, and Manchester the brand become interchangeable. In its need to communicate a designerly 'up-to-the-minuteness' as the city's heritage, an exchange of meaning takes place between Manchester the brand and Saville the creative. Saville's hand is deployed across the city's designscape, not just through the Urbis exhibition but, for example, through his historical association with Factory Records, to inflect this tradition of modernity with the desired notions of 'attitude' and 'edge'.

Hemisphere undertook and commissioned extensive research in developing the brand strategy. This included primary research with residents, businesses and visitors, a "reputation audit with key opinion formers", consultation with 92 "stakeholders, council members and members of Manchester's creative community" (Hemisphere, 2003, p. 2). Despite this impressive roster, a notable self-referentiality is evident in parts of its strategy in terms of the milieu the consultancy itself inhabits and how this was written into the brand strategy. The clue to this is in its expansion on its 'Respect' brand value. The report continues

> Market Street may be the economic engine of Manchester's retail scene but it is the independent and quirky Northern Quarter that is seen as most epitomising 'Manchesterness' (Hemisphere, 2003, p. 9).

Manchester's city-centre creative cluster, the Northern Quarter has evolved as both a site of design production and consumption, with its rich network of creative businesses supporting and supported by a designerly leisure infrastructure of bars and cafés (see Wansborough and Mageean, 2000). Hence it contains all the features of an urban designscape. The habitus it fosters may be very different from Barcelona's. It is not characterised by a core upper-middle-class élite to champion a refined modernity, but by a clustering of entrepreneurial creative workers spanning a range of aesthetic and operational practices. While shaping stuff for sale or use is the core of its business, it also exists, to a degree, as the hub of cultural planning and identity for the city. Hemisphere (who unsurprisingly also has its own offices within the Northern Quarter) claims a central role for the Northern Quarter's milieu in the shaping of an urban identity for Manchester.

A more extended enquiry into the symbolic role of creative quarters in anchoring the cultural capital of a location is considered in a separate paper (Julier, 2004). But it is clear that the Northern Quarter is appropriated and corralled as a signifier for 'Manchesterness' within an official version of the city. 'Manchesterness' is mediated through design outcomes but also by design practice itself. Its network sociality, its louche avant-gardism, its inhabiting of liminal spaces are made to symbolise that identity. Thus, the Manchester case ultimately exhibits an attempt to produce a more programmatically defined identity as a pre-existing designscape is articulated into an officially sanctioned marketing ploy. In shifting the city's brand towards this identity, an attitudinal marker is layed down that mirrors Manchester's conversion from municipal socialism to an entrepreneurial location, backed by a political shift towards 'élite consensus' (see Quilley, 1999, 2000). A network of interests links a designerly élite (The McEnroe Group), a symbolic location (the Northern Quarter), a key designer figure (Saville) and the city council's government marketing strategy to produce a particular version of Manchester for local and global consumption. Specific sites of public consumption, such as the Northern Quarter and Urbis, and everyday practice help to mediate these values.

The City of Hull: An Aspirational Habitus

What happens, however, when very little aesthetic value exists that can be appropriated and corralled into a set of brand-values? Can a location with a much more limited design production and consumption infrastructure or identity create its own kind of designscape? We have seen an example of a culturally rich city where a specific but largely unplanned identity of regeneration emerged (Barcelona). We have seen how a very specific version of a city's creative resources is appropriated into planned marketing of place-identity (Manchester). The City of Hull presents a reverse challenge in that a planned and articulated identity is mobilised, in effect, to 'jump-start' a system of cultural capital for the city.

Hull, a northern UK city with a population around 250 000, had suffered severe deindustrialisation from the 1980s with the loss of its shipbuilding and fishing industries and was blighted by its drab reputation in comparison with other UK centres. During mid 1990, informal discussions began among its city leaders as to the problem that Hull had in its poor external image. As a result, Wolff Olins were commissioned in 1997 to review Hull's image—then a highly unusual move on the part of a Labour-controlled local authority—which they did by a visual audit of the city and, following this, a brand strategy for the city. Wolff Olins, the London-based multinational design and branding consultancy, were at the time building a considerable reputation in place-branding (see Julier, 2000, ch. 6). Wolff Olins' findings were that the city suffered from a lack of consistency and coherence in its identity, as much in its design details—signage, city literature, architecture—as in the way civic interests organised and communicated themselves (D. Tate, Head of Marketing, CityImage, personal interview 1 September 2004).

The resulting brand that Wolff Olins developed stemmed from the notion of 'Pioneering Hull', itself partly drawn a recognition of the city's historical reputation in scientific discovery. From this flowed the five brand values of 'challenging', 'discovering', 'creating', 'innovating' and 'leading' (Pywell and Scott, 2001). However, while a brand promise was in place, Wolff Olins' key finding was the lack of aspiration and confidence of Hull's population. Thus it developed the 'Top Ten' goal for the city as an objective for all areas of everyday life within the city. The aim was to engender this notion into the everyday parlance of Hull's citizens and institutions.

This brand was taken further into the creation of the city's private–public sector Local Strategic Partnerships in 1998. Initiatives under the unified heading of 'CityVision' were created—CityEducation, CityEconomy, CityHealth and so on. City-Image was established to manage and develop the implementation of the brand and image-related issues emanating from this. Through a bond system, local companies bought stakes in this image enhancement programme, the money for which was used to lever further funds from the UK government and the European Union for urban regeneration programmes. This process also engendered a united front on the part of corporate and political élites in selling the city (Davies, 2003).

This brand strategy evolved in concert with a regeneration strategy for the city rather than, as in most city cases, a tourism strategy. It has been implemented, for example, through targeted communications towards the attraction of greater levels of external funding and investment on the one hand, and 'spreading the word' within the city on the other. As a result, CityImage has been anxious to avoid any use of the traditional jargon of marketing and branding in its activities. Conscious that its primary audience is both Hull's city leaders and its rank-and-file population, the 'Top Ten' aspiration had to be normalised and articulated within the terms of these stake-holders rather than by employing the special-ist languages associated with, for instance, advertising or product launches. Implemen-tation of the brand was not 'campaign led'. There was no advertising budget or straplines and slogans (D. Tate, Head of Marketing, CityImage, personal interview, 1 September 2004).

The Wolff Olins brand scheme for Hull was developed and managed in order to overturn aspirations of Hull. In doing so, it attempted to provide an aspirational narrative for every-day practice and a new vector in the urban

habitus. On a highly visible level, it was directly responsible for the architecture of two significant buildings. Following a Wolff Olins presentation of its brand strategy to city leaders, architecture was 'recommis-sioned' in 1999 for the Arco National Redis-tribution Centre in Hull. It went on to win a national Civic Trust Commendation for its design in 2002. The Wolff Olins presentation also led to Terry Farrell and Partners being commissioned to design 'The Deep', £45 million visitor attraction (Tate, 2004). Farrell subsequently developed a masterplan for the redevelopment of the city centre. Key figures have thus been directly influenced by the programme which in turn has impacted on the built environmental quality of the city.

Without an extensive qualitative survey of attitudes and practices of Hull's population, it is impossible to demonstrate empirically the end-effect of this programme in terms of ordinary lives. However, it is clear that it has been developed and implemented not just to lay down atittudinal markers of the city, but to adjust its own sense of cultural capital.

We have seen a clear relationship between the aesthetic dispositions of an élite, collective consumer aspirations (beyond, merely, the built environment) and urban identity forma-tion in Barcelona. In the case of Manchester, a similar dynamic has been more self-consciously experimented with. The transfer-ability of this analysis to other locations is of course tempered by the specific contextual circumstance of each one. However, the concept of various 'urban designscapes', par-ticular to distinct locations and historical moments, is worth pursuing, analysing and critiquing.

Conclusion

By conceptualising urban agglomerations in terms of designscapes, we are able to reach beyond the superficiliaties of branding or built form in our pursuit of an understanding of the role of culture in urban regeneration and identity formation. Here, prominence is given to the dynamics of taste formation through an appreciation of the roles of the multiple actors and locations within this process. It is necessary to trace the flows of aesthetic information between stakeholders while bearing in mind the possibility that these carry specific values and meanings that are framed by specific élites of respective locations. These élites create the narrations by which artefacts are interpreted and articulated. These narrations link the meanings of both private and public design consumption within the metropolis and, indeed, provide cues for the performance of everyday life.

An understanding of designscapes reveals them as more haphazard than branding theorists would perhaps like. The recognition that, as we have already seen, any development of place-identity is one of nurturing pre-existing information resonates with the possibility that this is a process of appropriation and reappropriation rather than invention. This is both slow-moving and swift. On the one hand, given the multiciplicity of stakeholders that constitute a designscape, it takes time for coherent messages to be formed and moved through its system. Equally, they can lose their coherence as they are received, used or even contested by its various nodes. On the other hand, while traces of urban heritage exist within these messages, they are principally about modernity, avant-gardism and the symbolic capital of creativity. Hence this requires the constant harnessing of an image of 'up-to-the-minuteness' and therefore the rapid appropriation of creative activities as signifiers of this image.

We have seen the ubiquity with which messages of 'vibrancy' and 'cosmopolitanism' are deployed across global cities. Beyond these, it seems that the élites that are firmly embedded within creative industries will forge more nuanced identities within these broad categories. These identities are nonetheless contingent upon their effective circulation through their respective designscapes for them to have relevance and interest to a wider population. For many, life continues untouched by these ambitions. For others, a new and different enthusiasm begins.

References

ANHOLT, S. (2002) Foreword, *Journal of Brand Management*, 9(4/5), pp. 229–240.

ASHWORTH, G. and GOODALL, B. (Eds) (1990) *Marketing Tourism Places*. London: Routledge.

BELL, M. and JAYNE, M. (2003) 'Design-led' urban regeneration: a critical perspective, *Local Economy*, 18(2), pp. 121–134.

BENNETT, R. and SAVANI, S. (2003) The rebranding of city places: an international comparative investigation, *International Public Management Review*, 4(2), pp. 70–87.

BHABA, H. (1990) Introduction, in: H. BHABA (Ed.) *Nation and Narration*, pp. 1–9. London: Routledge.

BITNER, M. J. (1992) Servicescapes: the impact of physical surroundings on customer and employees, *Journal of Marketing*, 56, pp. 57–72.

BOHIGAS, O. (1989) *Combat d'Incerteses: Dietari de Record*. Barcelona: Edicions 62.

BUTLER, J. (1990) *Gender Trouble*. London: Routledge.

CALAVITA, N. and FERRER, A. (2000) Behind Barcelona's success story: citizen movements and planners' power, *Journal of Urban History*, 6, pp. 793–807.

CAMPBELL, C. (1987) *The Romantic Ethic and the Spirit of Modern Consumerism*. Oxford: Basil Blackwell.

CAMPI, I. (2003) De la industria sin diseño al diseño sin industria, *La Vanguardia: Culturas* (Supplement), 8 October, pp. 4–5.

CAROL, M. (1988) *15 Años de BCD*. Barcelona: Barcelona Centro de Diseño.

CHATTERTON, P. and HOLLANDS, R. (2003) *Urban Nightscapes, Youth Cultures, Pleasure Spaces and Corporate Power*. London: Routledge.

CRILLEY, D. (1993) Architecture as advertising: constructing the image of redevelopment, in: G. KEARNS and C. PHILO (Eds) *Selling Places: The City as Cultural Capital, Past and Present*, pp. 231–252. Oxford: Pergamon Press.

DAVIES, J. (2003) Partnerships versus regimes: why regime theory cannot explain urban coalitions in the UK, *Journal of Urban Affairs*, 25(3), pp. 253–269.

Design Week (2004) Remoulding Manchester, 8 April.

EVANS, G. (2003) Hard-branding the cultural city—from Prado to Prada, *International Journal of Urban and Regional Research*, 27(2), pp. 417–441.

FEATHERSTONE, M. (1991) *Consumer Culture and Postmodernism*. London: Sage.

FLORIDA, R. (2002) *The Rise of the Creative Class*. New York: Basic Books.

FRAMPTON, K. (1986) Place-form and cultural identity, in: J. THACKARA (Ed.) *Design after*

Modernism: Beyond the Object, pp. 51–66. London: Thames & Hudson.

GENERALITAT DE CATALUNYA (1984) *Llibre Blanc del Disseny. Vol. 1 Disseny Industrial; Vol. 2 Disseny Gràfic; Vol. 3 Disseny Artesà*. Barcelona: Generalitat de Catalunya, Servei Central de Publicacions.

HAJER, M. A. (1993) Rotterdam: re-designing the pubic domain, in: F. BIANCHINI and M. PARKINSON (Eds) *Cultural Policy and Urban Regeneration: The West European Experience*, pp. 48–72. Manchester: Manchester University Press.

HEBDIGE, D. (1988) *Hiding in the Light: On Images and Things*. London: Comedia.

HEMISPHERE DESIGN AND MARKETING CONSULTANTS (2003) *A brand strategy for Manchester: Stage 1 Summary: Brand Values & Positioning*. Hemisphere Design.

HUBBARD, P. (1996) Urban design and city regeneration: social representations of entrepreneurial landscapes, *Urban Studies*, 33(8), pp. 1441–1461.

JULIER, G. (1990) Radical modernism in contemporary Spanish design, in: P. GREENHALGH (Ed.) *Modernism in Design*, pp. 204–223. London: Reaktion.

JULIER, G. (1995) *Design institutions and the transition to democracy: a comparative case study of Spain and Hungary*. Paper given at 'Design, Government Initiatives and Industry' International Conference, Brighton University.

JULIER, G. (2000) *The Culture of Design*. London: Sage.

JULIER, G. (2004) *Design, creative quarters and New labour*. Paper given at 'The Politics of Design', Design History Society Annual Conference, Belfast (unpublished).

JULIER, G. and NAROTZKY, V. (1998) *The redundancy of design history*. Paper given at 'Practically Speaking' Conference, Wolverhampton University, December (http://www.leedsmet.ac.uk/hen/aad/artdesresearch/papers_redundancy.htm).

KEARNS, G. and PHILO, C. (Eds) (1993) *Selling Places: The City as Cultural Capital, Past and Present*. Oxford: Pergamon Press.

KOPYTOFF, I. (1986) The cultural biography of things: commoditization as process, in: A. APPADURAI (Ed.) *The Social Life of Things: Commodites in Cultural Perspective*, pp. 64–94. Cambridge: Cambridge University Press.

KOTLER, P. and GERTNER, D. (2002) Country as brand, products, and beyond: a place marketing and brand management perspective, *Journal of Brand Management*, 9(4/5), pp. 249–262.

KOTLER, P., HAIDER, D. and REIN, I. (2002) *Marketing Places: Attracting Investment, Industry and Tourism to Cities, States and Nations*. New York: Free Press.

LASH, S. (2002) *Critique of Information*. London: Sage.

LASH, S. and URRY, J. (1994) *Economies of Signs and Spaces*. London: Sage.

LEACH, N. (2002) Belonging: towards a theory of identification with space, in: J. HILLIER and E. ROOKSBY (Eds) *Habitus: A Sense of Place*, pp. 281–298. Ashgate: Aldershot.

MANCHESTER CITY COUNCIL (2004) Manchester appoints creative director, *Press Release*, 17 March.

MARKETING MANCHESTER (2002) *Manchester Primer handbook* (www.manchesterprimer.co.uk).

MARSHALL, T. (2004) *Transforming Barcelona*. London: Routledge.

MCNEILL, D. (2003) Mapping the European Left: the Barcelona experience, *Antipode*, 35(1), pp. 74–94.

MCROBBIE, A. (1998) *British Fashion Design: Rag Trade or Image Industry?* London: Sage.

MCROBBIE, A. (2002) Clubs to companies: notes on the decline of political culture in speeded up creative worlds, *Cultural Studies*, 16(4), pp. 516–531.

MOLOTCH, H. (1996) L.A. as design product: how art works in a regional economy, in: A. J. SCOTT and E. W. SOJA (Eds) *The City: Los Angeles and Urban Theory at the End of the Twentieth Century*, pp. 225–275. Berkeley, CA: University of California Press.

MOMMAAS, H. (2002) City branding: the necessity of socio-cultural goals, in: ONDER REDACTIE VAN URBAN Affairs and V. PATTEEUW (Eds) *City Branding: Image Building and Building Images*, pp. 32–47. Rotterdam: NAi Publishers.

MORGAN, N., PRITCHARD, A. and PRIDE, R. (2002) *Destination Branding: Creating the Unique Destination Proposition*. Oxford: Butterworth-Heinemann.

M.R. (1998) Soft branding works for summit members, *Lodging Today*, July, p. 22.

MUÑIZ, A. and O'GUINN, T. (2001) Brand community, *Journal of Consumer Research*, 27, pp. 412–432.

NAROTZKY, V. (2000) 'A different and new refinement': design in Barcelona 1960–1990, *Journal of Design History*, 13(3), pp. 227–243.

NIXON, S. (2003) *Advertising Cultures*. London: Sage.

OLINS, W. (1999) *Trading Identities: Why Countries and Companies Are Taking On Each Others' Roles*. London: The Foreign Policy Centre.

PAPADOPOULOS, N. and HESLOP, L. (2002) Country equity and country branding: problems and prospects, *Journal of Brand Management*, 9(4/5), pp. 294–315.

POYNOR, R. (2003) Different colour, different shades, in: E. KING (Ed.) *Designed by Peter Saville*, pp. 155–158. London: Frieze.

PYWELL, J. and SCOTT, L. (2001) Pioneering spirit, *Locum Destination Review*, 3, pp. 30–32.

QUILLEY, S. (1999) Entrepreneurial Manchester: the genesis of an elite consensus, *Antipode*, 31(2), pp. 185–211.

QUILLEY, S. (2000) Manchester first: from municipal socialism to the entrepreneurial city, *International Journal of Urban and Regional Research*, 24(3), pp. 601–615.

RICARD, A. (2003) *En Resumen* ... Barcelona: Angle Editorial.

SCASE, R. and DAVIS, H. (2000) *Managing Creativity: The Dynamics of Work and Organization.* Milton Keynes: Open University Press.

SHAUGHNESSY, A. (2004) Peter Saville—Creative Director of Manchester, *Voice: AIGA Journal of Design and Professional Practices*, 6 May.

TSO, J. (1999) Do you dig up dinosaur bones? Anthropology, business, and design, *Design Management Journal*, 10(4), pp. 69–74.

URBAN TASK FORCE (2000) *Our Towns and Cities: The Future: Full Report.* London: Office of the Deputy Prime Minister.

WANSBOROUGH, M. and MAGEEAN, A. (2000) The role of urban design in cultural regeneration, *Journal of Urban Design*, 5(2), pp. 181–197.

WARD, S. (1998) *Selling Places the Marketing and Promotion of Towns and Cities, 1850–2000.* London: E & FN Spon.

ZUKIN, S. (1991) *Landscapes of Power: From Detroit to Disney World.* Berkeley, CA: University of California Press.

ZUKIN, S. (1995) *The Cultures of Cities.* Cambridge, MA: Blackwell.

Interruptions: Testing the Rhetoric of Culturally Led Urban Development

Malcolm Miles

Introduction

Since the 1980s, in the UK and US, the arts (within a broader category of the cultural and creative industries) have gained a key role in strategies to deal with urban problems from social exclusion to the rehabilitation of post-industrial sites. Persuasive advocacy on the part of professionals and organisations within the cultural industries has been a factor in persuading governments that, in post-industrial situations, the cultural industries, and related knowledge sector of electronic communication and higher education, can provide a new economic base. A small number of successful cases tends to be advanced as evidence that a cultural turn in policies for urban renewal can deliver revitalisation of post-industrial cities. These cases often centre on a new flagship cultural institution. Examples include Tate Modern in London and the Guggenheim in Bilbao. In

other cases, an entire district may be redesignated as a cultural quarter. Examples include the Rope Walks Quarter in Liverpool and El Raval in Barcelona. The latter also includes a flagship venue, the Museum of Contemporary Art, Barcelona (MACBA). Apart from drawing visitors into an area, such venues and recodings of a district tend to encourage a proliferation of small, broadly cultural businesses, from graphic design and architectural design firms to designer-bars and boutiques, all catering for a new cultural class, as it were.

Following the turn in Glasgow's fortunes after being a European Capital of Culture in 1990, it is easy to understand how city authorities and developers alike are captivated by cultural projects. As Sharon Zukin notes: "so much of the dominant capitalist economy has ... undergone a cultural turn" (Zukin, 1996, p. 226). The allure of redeveloped waterfront sites such as Baltimore's harbour district, London's docklands or Cardiff Bay provides an aesthetic gloss for commercial schemes which frequently erase all traces of the histories of their sites. But there are difficulties. The recoding of a district as a cultural quarter may lead to gentrification—a shift from multiple to single occupancy and from rent to owner-occupation of housing being a key aspect of this—and a marginalisation (or peripheralisation) of dwellers who become constituted as a residual public. A cultural zone can easily be read as a zone of affluence. But setting that contentious (and complex) issue aside, there are still questions: is advocacy for the creative industries itself creative (like creative accounting)? And to what extent can policies and strategies which are successful in one city be mapped onto others in which conditions differ? As John Myerscough (1988) argues, Ipswich lacks the grounded cultural infrastructure of Glasgow and would not produce a similar return on the same culturally based investment. Similarly, while Barcelona is advanced as a model of a successful cultural economy (Degen, 2004), it remains exceptional on account of a unique cultural heritage and the political impetus to revive Catalan culture after the end of the Franco period, quite apart from a policy to attract cultural rather than mass-market tourism from the early 1990s onwards (Dodd, 1999). To see culture, with its legacy in liberal reformism of a universal benefit, as a universal solution to present-day urban problems may, then, be romantic.

I wonder, too, whether advocacy for the cultural industries as general problem-solvers is based on more than vested interest, or may represent a co-option of culture to the agenda of marketisation. My first aim, then, in this paper, is to reflect on the cultural turn in urban redevelopment. I start by reconsidering this cultural turn, asking what is meant by culture, examining aspects of culturally led urban redevelopment and asking whether Adorno's use of the term 'the culture industry' to denote an industry of mass deception remains valid. But my second aim—because I think cynicism is required in the present political and cultural climate, but is not by itself enough—is to ask how cultural producers such as artists might contribute critically to processes of urban change. I do this by looking at a small number of art projects which do not subscribe to the dominant agenda of culturally led urban redevelopment, but operate in urban settings. These are not widely known and tend to be absent in art journals (which foreground art in gallery spaces). Such projects may, however, offer an alternative form of cultural intervention to conventional public art.

Voices of Dissent

I am encouraged in this investigation by criticality on the parts of some artists working in urban situations today. This is the more interesting because many artists have gained material benefit from the cultural turn in urban policy through commissions for public art or the provision of new gallery spaces in which to exhibit their work. Among the dissenting voices are those of Sarah Carrington and Sophie Hope (the artist group B + B)

> In the UK the integration of art into the Government's policy on social inclusion and regeneration relies heavily on utopian notions of art as an empowering tool ... The 'art pill' is now dished out by New Labour in an attempt to empower and effect change through the participatory values of art (www.welcomebb.org.uk).

George Yúdice observes a similar situation in the US

> No longer restricted solely to the sanctioned arenas of culture, the arts would be literally suffused throughout the civic structure, finding a home in a variety of community service and economic development activities—from youth programs and crime prevention to job training and race relations—far afield from the traditional aesthetic functions of the arts (Larson, 1997, pp. 127–128; cited in Yúdice, 2003, p. 11).

Yúdice points out that this is a defensive posture after attacks on the arts in Congress and threats to the survival of the National Endowment for the Arts (NEA), from whom the above was drafted. Multifunctionality in the arts mirrors multitasking in post-industrial patterns of flexible employment (see Sennett, 1998) and seems to be led by the arts funding system. B + B note complexities which interrupt the scenario of culturally led redevelopment. They argue that art's aesthetic distancing of reality offers either a critical space, or a resignation to the world as given (the better future reassigned to an aesthetic dreamland equivalent to Heaven); and that art subsumed to the agenda of a regime dedicated to marketisation will not retain a radical edge. Perhaps the art pill is a placebo. Perhaps the rhetoric is hollow.

My concern echoes Monica Degen's in her critique of the aestheticisation of spaces in Barcelona and Castlefield, Manchester (2002). And as a cautionary tale, I would cite the rebranding of the UK as Cool Britannia by New Labour after the 1997 election victory and the failure of the Millennium Dome (which I take in some ways as an emblem of Cool Britannia). The Dome and Tate Modern, while contrasting cases in terms of visitor numbers, are not far apart on the south bank of the Thames. Both are flagship cultural sites. Tate Modern has exceeded expectations of visitor numbers to become a key node of metropolitan cultural life, while the Dome attracted only a fraction of its intended numbers despite insistent media promotion and special offers on ticket prices. Tate Modern, of course, is free—although its café, restaurant and bookshop are not. Yet the oddest thing seems that Tate Modern has a collection of modern art, which is a specialist interest and for most an acquired taste, while the Dome attempted to display the supposedly universal qualities of the nation's values, spirituality, diversity and technological achievements. Tate has moved the cultural centre of London across the river (aided by a new bridge) but has done so less by converting the city's diverse publics to modern art than by becoming a new social space, a place to

meet, eat, buy books and be seen. Tate is cool, and this results from a marketing exercise, but through an understatement which enables its publics to entertain the notion they have produced that cool themselves. It is the success of Tate Modern not the failure of the Dome which continues to attract public authorities and private developers alike to the strategy of culturally led urban redevelopment. But it may be the Dome which better represents the rhetoric of a universalised cultural intervention in national and urban life.

It may be also that interventions of a more specific and localised kind, as I read in the art projects I cite in the second part of the paper, have a greater resonance for those who encounter or participate in them. If so, perhaps for at least a few people, cultural work is a means to a new approach to the many problems which beset city dwelling in a post-industrial period.

1. Urban Culture and Policy

In this section, I begin by noting the ambiguities which attend the term culture. This begs the question as to what and in particular whose culture is utilised in urban redevelopment. I then examine the vicissitudes of culturally led urban redevelopment since the 1980s. Finally in this section, seeking an appropriate theoretical framework, I reconsider Adorno's argument on the culture industry as an industry of mass deception.

1.1 Culture and Cultures

There are ambiguities in the use of the term culture in strategies for culturally led urban renewal. Beginning *The Cultures of Cities* with a reference to cities as centres of culture, Sharon Zukin rehearses some of the term's uses

The Acropolis of the urban art museum or concert hall, the trendy art gallery and café, restaurants that fuse ethic traditions into culinary logos—cultural activities are supposed to lift us out of the mire of our everyday lives and into the sacred spaces of ritualized pleasures (Zukin, 1995, p. 1).

The list implies a series of social strata more complex than a set of social classes: a high art élite whose cultural capital may not equate with money capital; new bohemians, yuppies who like rubbing shoulders with artists and social others in cultural spaces; consumers of ethnic foods, who favour hybrid cuisines. Yet all are metropolitan types of some affluence. Absent from the list are those without access to consumption. Zukin writes, still, from a concern for social justice and her use of the term 'cultures' (rather than 'culture') in her title indicates a concern for the everyday ways of life which constitute culture in an anthropological context. This recognition of the ordinary informed the inception of cultural studies as an academic discipline in the 1960s and Zukin cites Raymond Williams' *The Country and the City* (1973).

Zukin's analysis of culturally led redevelopment in New York City affirms that culture in redevelopment tends to be Culture—the traditional high arts of the Metropolitan Museum or the avant-gardism of the Museum of Modern Art. This, she points out, does not mean that artists embrace the agenda of culturally led redevelopment, although their production feeds an industry in which the insertion of new cultural spaces raises the value of surrounding real estate while museum boards of trustees offer networking opportunities to developers. Zukin remarks that the cultural labour force depends on a range of ways to earn a living—making art and washing up in bars—and that many in the arts "are supposed to live on the margins ... used to deprivation" (Zukin, 1995, pp. 12–13). The growth of an arts infrastructure in the past two decades, and in the UK a significant increase in public funding for arts management since the 1990s, may have taken cultural managers into a more affluent lifestyle—able to eat in the restaurants which denote a cultural zone (in which it is more likely that immigrants rather than artists will wash dishes)—but a majority of artists who do not

gain international reputations remain in a marginal economic category. But the culture of the cultured class is cultivated; it is like the cultivation of taste in the 18th century; it is equally a way of life expressing the value of culture (or culture as a value) in acts of cultural consumption which extend beyond the visual and performing arts to design and architecture, new media, food and drink, fashion and modes of transport.

What emerges is a meaning of culture specific to the spaces of post-industrial urban redevelopment. It bridges the anthropological (a way of life or, more carefully stated, a set of habits of everyday living which express and articulate a set of values) and the aesthetic (the arts and their appreciation by suitably educated minds). It is the culture of a class, diverse in background but with a disposable income, which uses cultural spaces. Walter Benjamin (1999) noted that the Paris arcades similarly housed a new class of window-shoppers and observers of others observing themselves.

1.2 Culture in Redevelopment: Strategies and Attractions

Part of Zukin's purpose is to draw attention to contestable and conflictual aspects of culture in a climate in which city authorities compete globally to rebrand their cities, in which image is all. She remarks that while culture offers ways to deal with difference it also "offers a coded means of discrimination, an undertone to the dominant discourse of democratisation" when styles which emerge at street-level—ripped jeans, for instance—"are cycled through mass media, especially fashion and 'urban music' magazines ... where, divorced from their social context, they become images of cool" (Zukin, 1995, p. 9). I share Zukin's concern that "The cacophony of demands for justice is translated into a coherent demand for jeans" (p. 9). Recent advertising by Nike confirms the potential for marginalisation to be subsumed in consumption when urban basketball spaces, in Berlin as in New York, are signed by what look like municipal notices but are in fact elements in an advertising campaign relying on recognition by a target public (Goldman and Papson, 1998; von Borries, 2003). But if cultural consumption is a means to defuse dissent—Zukin comments that "culture is also a powerful means of controlling cities" (Zukin, 1995, p. 1)—for city authorities it may be more a competitive edge in a campaign for inward investment. By commissioning a highly visible piece of public art or employing an internationally recognised architect, a city may purchase a place on a notional international culture map. The decision by Barcelona's city authorities to site a World Trade Centre designed by I M Pei on its redeveloped waterfront denotes such intention to be a world city.

The specifics vary, but culturally led urban redevelopment tends to include the following: the insertion of a flagship cultural institution in a post-industrial zone, often a waterfront site, to lever private-sector investment in the surrounding area and attract tourism; the designation of a neighbourhood as a cultural industries quarter for small- and medium-size businesses in the arts, media and leisure. The definition of what qualifies as a cultural industry varies, Allen Scott (2000) taking the broadest approach to include furniture manufacture, leather, perfume and other commodities alongside the arts and film in cities such as Los Angeles and Paris. For Myerscough (1988), the focus was more narrowly on the visual and performing arts and heritage; and for Charles Landry and Franco Bianchini (1995), it is the arts, media and cultural consumption which contribute to a creative city. Given that culturally led redevelopment occurs in deindustrialised conditions, it is not surprising that outposts of cultural recoding are geographically juxtaposed with areas of residual deprivation. Graeme Evans (2004, p. 71) notes that Hoxton in London, an area of multiple deprivation for many inhabitants, is also "the capital's trendiest area". Like New York's SoHo in the 1980s, it combines a cluster of arts and media venues, including the White Cube Gallery in which the work of artists such as Antony Gormley is shown, with increasingly

high-rent apartments. Evans remarks that the cultural industry quarter models currently promoted in urban regeneration "tend to neglect both the historical precedents and the symbolic importance and value of place and space" (Evans, 2004, p. 91). Citing Landry (2000), he argues elsewhere that the vogue for culture does not include broad representation of cultural producers or communities, which "mirrors the professionalisation and bureaucratisation of both cultural and other public policy realms and decision-making structures" (Evans, 2001, p. 277). The outcome is a growth in cultural infrastructure but not in support for cultural producers such as artists, writers and performers. Members of those groups may individually gain from the provision of new venues, but this does not in itself support the experimental, non-market-led production of new work. Further, many new cultural buildings in the UK were allocated capital funding from the national lottery on the basis of projected visitor numbers supplied by cultural industries consultants, but not revenue funding. One casualty of this system was the Earth Centre in Doncaster, on the site of a redundant colliery and employing redundant miners, which closed when visitor numbers failed to match targets which may have been unrealistic.

It seems that Tate Modern can capitalise on its occupation of a redundant industrial site, its ex-industrial building having "a fashionably squatted aspect" (Leslie, 2001, p. 3), but at the Earth Centre an imaginative response to environmental sustainability, with reed-bed water cleansing and gardens for arid places, fails in a less glamorous location more than two hours by rail from London. Tate Modern responded to employment needs in Southwark, one of London's poorest boroughs, through a training programme for local people which more or less guaranteed them an interview for jobs in security and catering, but the lesson remains that, first, capital projects do not always survive; and, secondly, although catering is the largest sector in Tate's employment it does not admit the operatives to the cultural realm or affluence of which the museum is an emblem. Failures such as that of the Earth Centre do not inhibit cultural industries advocates, any more than a lack of evaluation of benefits inhibited the commissioning of public art in the 1980s and early 1990s. According to Sara Selwood, in a book ironically titled *The Benefits of Public Art* (1995), the claims made for commissions and projects were largely so vague as to be undemonstrable. Public art gained from the fluency of its advocates—the arts breed effective verbal communication—and a willingness of government through the Arts Council to back campaigns for schemes such as Per cent for Art (through which a percentage, usually one, of capital budgets would be reserved for art and craft commissions or projects) and was in some ways a dry run for the success of cultural industries advocacy (Landry, 2000; Landry and Bianchini, 1995) in persuading municipal authorities to adopt notions of a creative city. There is a tendency to generalisation in statements such as

> In a number of American [sic] cities, leading strategists of 'downtown rejuvenation' have argued that arts-led investment is the most efficient way of beginning the process of raising morale and developing 'atmosphere' in ... low-status and moribund districts (Bianchini *et al.*, 1988, p. 14)

and to aestheticisation as in "Urban design is essentially about knitting together different parts of the city into a coherent artefact" (Landry and Bianchini, 1995, p. 28). This might be set beside Jane Jacob's remark that "*a city cannot be a work of art*" (Jacobs, 1961, p. 373; original emphasis).

1.3 Contestable Terms

A difficulty is that meanings of culture as the arts, a way of life and means of a symbolic economy may be fused as if they denote a unified concept. A further difficulty is that cultural policy tends to remain instrumentalist despite the insights of complexity theory (Byrne, 1997; Cilliers, 1998) that outcomes of a given intervention cannot be predicted

in the way assumed in the rational planning model of the inter- and post-war periods, but will be affected by even minor shifts in conditions. This adds to the difficulty of mapping solutions from one city to another. A third difficulty is that the benefits of cultural redevelopment are unevenly distributed. I accept that cultural flagships can contribute to more confident perceptions of a city, including by some of its publics as well as investors or tourists. Writing on Glasgow in 1990, Peter Booth and Robin Boyle say that the opening of a new gallery for the Burrell Collection "was undoubtedly the catalyst that drew the different components together" (Booth and Boyle, 1993, p. 31). In the UK, cases such as Tate Modern, Tate of the North in Liverpool (to be European City of Culture in 2008), the Lowry Museum in Salford and the Baltic in Gateshead (with the nearby Sage centre for chamber music) can be advanced as having changed external perceptions of their sites. But, as the arts have moved in the UK from being administered as a public service to being managed as businesses paying their way in increased property values, job creation and tourism, so what is sometimes called urban regeneration (with an implication of community benefit) has become urban redevelopment. Rosalyn Deutsche (1996, pp. 49–109) notes the uneven benefits of redevelopment in New York when homeless people, some evicted as an outcome of gentrification, were cleared from spaces such as Grand Central Station by Mayor Koch. The art displayed in iconic cultural spaces meanwhile becomes emblematic of a new affluence, this reading displacing readings based on its histories of production and reception which include histories of dissent and refusal of art's commodification. Esther Leslie writes

> Tate Modern is not just trendy, but in the vanguard of a reinvention of cultural spaces worldwide ... the expertise of art workers is leased out to business and education, with online gift shops, travel planning, digital reproductions for download and so on (Leslie, 2001, p. 3).

In an uncanny mix of élitism and populism, Tate retains a role of interpreting modern visual culture in the selectivity of its exhibits *and* portraying itself as free of the class aspect of museum-going. It remains, however, an instrument of liberal reform, in its early days at Millbank having offered an education in taste and behaviour for the lower classes (Taylor, 1993) and today providing free admission to a modernised cultural ambience which may be no less subject to codes of behaviour—although the value to which the code lends coherence might now be consumption rather than liberal education. The modernisation of Tate, which for Leslie is marketisation, masks possible arguments as to whose culture is promoted, in the context of complex cross currents of class, race, gender and cultural as well as money capital. In other cases, too, cross-currents complexify the situation: Booth and Boyle (1993, p. 40) cite criticisms of Glasgow's Capital of Culture programme, promoted by Saatchi and Saatchi, as marginalising the city's working-class culture and emphasising cultural tourism to the detriment of support for local cultural producers; and Julia Gonzalez (1993, pp. 84–86) notes a contest of values between the internationalism of Bilbao's élites represented by the Guggenheim and a grassroots interest in the arts as articulating Basque nationalism. While Disney stands for a depoliticised global culture and Guggenheim is an international art venue brand, it may be that local cultures are both more politicised and under threat. Joost Smiers (2003, p. 103) reports, for instance, that "Little is left of the Egyptian film industry" as a result of a mix of factors including censorship, a boycott of Egyptian films by some Arab states after the country's separate peace with Israel and "competition with American [sic] products". He adds that "By the end of the 1980s Brazilian cinema had been all but destroyed ... from the results of the hegemony of the neoliberal project" (Smiers, 2003, p. 103). Still, when cities are sites of increased migration and the pervasive discourse is of difference, cultural dialogue offers a potential ground for contestation and a non-confrontational means to maintain a creative tension within difference.

In the 1970s, Raymond Williams saw in popular culture

a complex argument about the relations between general human development and a particular way of life, and between both and the works and practices of art and intelligence (Williams, 1976, pp. 80–81).

In 1995, Landry and Bianchini stated that

Ethnic ghettoes are unlikely to contribute to solving the wider problems of cities [while] New ideas can be generated through cultural crossovers, as in the success of young British Asians who have synthesised 'bhangra' music (Landry and Bianchini, 1995, p. 25).

Justin O'Connor sees cultural intermediaries (entrepreneurs in popular music, for instance) playing key roles in the new economies of sites such as Castlefield (O'Connor and Wynne, 1996; O'Connor, 1998). In another way, also addressing current urban conditions, the Creative Partnerships scheme funded by the UK government links artists with schools in which they are seen as injecting a creative energy which will have general benefit to the education system. In both cases—difference and education—the arts are perceived as catalysts to the solution of social problems. Yúdice (2003) outlines a parallel scenario in the US. This is an extension of the argument used for public art in the 1980s and early 1990s (Shaw, 1991), that public art contributes to place identity and acts as a catalyst for economic recovery. This might reflect a view that social problems, such as street crime, are produced by deprivation; arts advocates are adept at annexing new social agendas and governments seem glad to accept this. (Cynically, I would say because the arts are cheaper and arts projects easier to understand than deeper enquiry into social problems.) But the result, in an ethos which requires auditable outcomes for public expenditure, is that the solutions favoured tend to be in the form of cultural *economies*—new areas of consumption trading on cultural identities—rather than a regeneration of local cultures.

New terms are inserted in the argument but lack specificity, denoting a lack of specific understandings and a hegemony of top–down strategies which displace grassroots tactics. The term 'vibrancy', for instance, is used by Andrew Kelly in Bristol's (failed) bid to be a European Capital of Culture in 2008

> For cities to develop, not just as regional centres, but on the national and inter-national scale, they need to become cultu-rally vibrant (Kelly, 2001, p. 16).

This is reminiscent of the Arts Council's *An Urban Renaissance*

> The arts are crowd-pullers. People find themselves drawn to places which are vibrant and alive (Arts Council, 1989).

The publication asserts that the arts are a catalyst for regeneration, a magnet for tourism and business, enhance the visual quality of a city's environment and provide a focus for community and individual develop-ment. This is a succinct summary of what has become a well-rehearsed case. Like much cultural advocacy, it cites references to 'American' models without looking to the specifics of evidence. Similarly, *City Centres, City Cultures*, produced for the Centre for Local Economic strategies (CLES) in 1988 claims that

> in many towns and cities the best strategic programme for improving the quality of life might well turn out to be based on developing a coherent and wide-ranging arts and cultural policy (Bianchini *et al.*, 1988, p. 10).

But if in the 1980s cultural provision still had something of a public-benefit ethos, in the 1990s the market-led approach of modern-isation has fed into it, so that Landry and Bianchini argue that "forming a well working public/private partnership is itself a creative act" (Landry and Bianchini, 1995, p. 51). Really?

1.4 Marketisation or Public Benefit?

Today's market emphasis contrasts with the public-benefit ethos of post-war liberalism in arts policy, even more with the social policy emphasis of the Greater London Council (GLC) in supporting community arts in the 1970s which extended the efforts of the Wilson government to democratise culture in the late 1960s. Jenny Lee, the first UK Arts Minister, echoed the concerns of working-class intellectuals like Raymond Williams and Richard Hoggart by promoting regional policies and a

> bridging of the gap between … the 'higher' forms of entertainment and the traditional sources—the brass band, the amateur concert party, the entertainer, the music hall and pop group (Jenny Lee; quoted in Willett, 1967, p. 203).

But determination of cultural norms tends to remain with an arts bureaucracy which reproduces an older parochialism, so that access is widened to a culture predetermined in the image of the governing cultural body. Arts publics are thereby rendered passive receivers of culture rather than being empowered to shape cultures. This liberal-reformist attitude persists in the findings of a panel chaired by architect Richard Burton in 1989–91 on the Per cent for Art policy. Burton's report seeks to make "contemporary arts and crafts more accessible to the public" and places "more interesting and attractive" (Shaw, 1991, p. 16) and relies on a univers-ality of cultural value. John Willett com-plained in 1967 that "The weakness of the policy of the last twenty years is that it rests on assumptions which are very seldom discussed" (Willett, 1967, p. 220) and perhaps something similar could be said now, even if the assumptions have moved from public benefit to globalised consump-tion. Marketisation, however, is clothed in a new rhetoric of social development in which the arts are utilised as representing non-commercial (aesthetic) value at the same time as taking on an increasing range of social issues. There was a precedent to an extent in the uses of community arts projects, such as painting murals on the gable ends of housing terraces, in the 1960s and 1970s, but perhaps then the publics for such projects were more specifically identified than the cultural tourists or investors of today's urban rebranding, and more involved in production of the work.

The extent to which the arts are now seen as problem-solvers is seen in the statistics for Single Regeneration Budgets (SRBs) in the UK. Of 66 SRBs in England in 1998–99, 31 included a cultural project; linked funding from bodies such as English Heritage and the environmental charity The Groundwork Trust brought the total support for cultural projects in SRBs that fiscal year to more than £100 million (Selwood, 2001, pp. 60–65). This is in the context of a rise in government annual funding for the arts to an average of £204.5 million in the years 1994–99 (Selwood, 2001, p. 183). Culture in SRBs, in other words, generated a budget in 1998/99 equivalent to about half that of the arts in the public sector (although more than £100 million was added to the latter from sponsorship). These figures can be seen also in context of £241.7 million of national lottery money distributed by the Arts Council and more than £300 million distributed by the Heritage Lottery Fund in capital schemes many of which contributed flagship projects to post-industrial urban redevelopment.

That the urban cultural turn is not always based on evidence for sustainable benefits to a city's publics is shown in studies of Birmingham's rebranding of its central business district around Centenary Square as a cultural zone; jobs were created, but tended to be temporary, part-time and low-paid (Loftman and Nevin, 1998). But rather than deconstruct the socioeconomic case for the arts, I want to draw attention to another difficulty: the privileging of the visual in culturally led urban redevelopment, again citing Bristol. This might seem a tangential argument, but I want to draw a parallel between this and the prevailing top–down approach of culturally led urban redevelopment.

1.5 Legibility or Interpretation from Above?

Bristol's bid to be a Capital of Culture exemplifies this. Kelly revives Kevin Lynch's (1960) concept of legibility in a campaign for Bristol Legible City. A booklet produced to explain the concept states

> Think of great cities and what makes them so distinctive, impressive and attractive. Without exception, the experience of the public realm—the quality of public spaces and the aesthetics of buildings and design—plays a huge part in shaping positive perceptions of a city (Bristol Legible City, 2001a).

Another interprets the aim through explanation of a new signage system across the city

> It's about building an identity for Bristol that can grow beyond signs to encompass everything from bus shelters and kiosks to street furniture and sculpture, becoming a symbol of a confident and successful European city (Bristol Legible City, 2001b, p. 13).

Here, a public art programme provides landmarks and Kelly states that the Legible City project brings together for the first time in a British city "a multidisciplinary team ... to consider the issue of city identity and legibility" (Kelly, 2001, p. 36). The new signs feature maps of the immediate environs, using a traditional bird's-eye viewpoint. It would have been interesting to see a street-level visualisation of routes through the urban landscape, even more so to draw attention to the multisensory aspects of a city's streetscapes—the smells, textures, sounds and so forth. But my point was that the legible city is a visual city and this involves a power-relation which Doreen Massey draws out in *Space, Place and Gender*

> It is now a well-established argument, from feminists but not only from feminists, that modernism both privileged vision over the other senses and established a *way* of seeing from the point of view of an authoritative, privileged, and male, position. The privileging of vision impoverishes us

through deprivation of other forms of sensory perception. ... But, and more important ... the reason for the privileging of vision is precisely its supposed detachment. Such detachment, of course, can have its advantages, but it is also necessarily a 'detached' view from a particular point of view. Detached does not here mean disinterested (Massey, 1994, p. 232).

This is more than a by-pass of Kantian aesthetics; it is the imposition of the dominant, masculine viewpoint of the conventional city plan which uses an ability to see-over (or oversee) as a metaphor for having power-over and was developed as a form of cartography following Alberti's invention of a device to map accurately a city's streets from a viewpoint on its circuit of walls in the 15th century, but is not the only possible means to map a city.

I want to take this a little further before returning to the main argument, not least because it emphasises the relation of urban representation (of which cultural rebranding is an aspect) to power. French Marxist philosopher Henri Lefebvre (1991) theorised that all societies have characteristic spatial practices—the perpendicular roads of a Roman city with a standardised siting of key functions on the axis throughout the Empire, for example—and that these are ideological. He then differentiated what he called conceptual (or representational) space from lived spaces (of representation). Conceptual space is epitomised by the architect's drawing, the town plan and the architectural metaphor used by Descartes when he writes, in his *Discourse* (1637), of an 'engineer' drawing regular places. Conceptual space is constituted by a unified, consistent and coherent system of signs—such as Cartesian co-ordinates—which, from another viewpoint, reduces the world to that system as if reality, if it exists, is only represented by it and cannot be directly experienced. In one way, conceptual space allows all kinds of operations which would otherwise be impossible, like planning a city and then building it according to a plan which the builders cannot see on the ground

(but according to which they are directed). It can also be argued that there is no possibility to articulate raw experiences, only to 'cook' them (to use a term from Levi-Strauss) through language. In another way, however, conceptual space marginalises what Lefebvre calls lived spaces (plural), the spaces of and around the body, of association and memory, of desire and hope, of shifting meanings, overlaid, as it were, on the spaces of buildings and streets, cutting at times through the grain of the vista. But Lefebvre (1991, pp. 78–79) is at pains to point out that lived spaces, even in Tuscany at the time in which perspective was invented, remain accessible to rural and urban dwellers. In extraordinary circumstances such as the toppling of the Vendôme Column during the Paris Commune of 1871, the reproduction of meanings which takes place (and precedence) in the routines of daily life and labour is interrupted by a production of new meanings in liberatory acts (which reenact a shift of power as in the destruction of a statue of the figurehead of the deposed power).

I would read the visual city, and the legible city, despite its progressive aspects as proposed by Lynch, as that produced in a universalised conceptual spatial realm and the city of multisense impressions, multiple and overlapping actualities, as produced in somatic spaces. Lefebvre does not see these as separate but as superimposed and complementary realms.

I am left asking why Kelly chose to base his campaign on Lynch more than 40 years after publication of *The Image of the City*. It may be that Kelly's approach reflects the assumptions of that era in other ways—notably in giving primacy to professionals rather than dwellers in the determination of his envisaged City of Culture. The arts advocacy of the 1980s, which sought the inclusion of art in the built environment and promoted artist-designed street furniture, likewise retained a profession-based model of urban change. There were proposed links between planners, architects, engineers, designers and artists, but limited recognition of the tacit expertise of dwellers on dwelling. If, in the 1960s, then, legibility was a progressive idea in dealing with the spaces between buildings which signature architecture ignored, it emphasised design and, for me, is out-dated by the 1990s literature of an architectural everyday informed by Lefebvre (Borden *et al.*, 1996; Harris and Berke, 1997; Cline, 1997; Hill, 1998; Wigglesworth and Till, 1998). Bristol Legible City sought to integrate the work of regional, national and international artists' and designers' in a long-term programme to "make the city easier to understand and more familiar" (Bristol Legible City, 2001b, p. 13) but the accent is on designing things rather than the informal and often invisible traces of occupation which constitute a familiarity of urban spaces for many dwellers. Looking around Bristol's cultural zone, the Harbourside area, today I find, crossing an artist-designed bridge, life-size bronzes of famous Bristolians Thomas Chatterton, William Penn and Henry Cabot, to whom Cary Grant was later added, in a theme park developed by a public–private finance initiative (perhaps the kind of creative partnership envisaged by Landry and Bianchini). These statues reproduce exclusions of race and gender in defining the public sphere; I doubt that doing away with plinths changes the power relations involved.

What does happen, familiarly, is that tourists photograph each other sitting (or walking for Cary Grant) next to the famous old men, hoping to find a little rubbed-off star-dust in the snapshots. In a not dissimilar way, claims for universal benefit in culture inform a tendency to universal solutions to urban problems, supposing that benediction is given from a position of power. This is despite departures from the conventional power relation of professional to dweller ('user', as Lefebvre critically employs the term) in radical planning (Sandercock, 1998) and the shift of allegiance from a public benefit to a market-led ethos in arts management and urban redevelopment. The cultural terrain remains all-encompassing in the scope attributed to it and continues to privilege the visual sense through a primacy of image and dominance of design professionals in the implementation of the images designed. Zukin sees such reductiveness "to a coherent visual representation" as a common element of culturally led redevelopment schemes, so that

> culture as a 'way of life' is incorporated into 'cultural products', i.e. ecological, historical, or architectural materials that can be displayed, interpreted, reproduced, and sold in a putatively universal repertoire of visual consumption (Zukin, 1996, p. 227).

And Bianchini writes that

> There are conflicts between . . . maintaining prestigious facilities for 'high' culture marketed to wealthy visitors which emphasize 'exclusiveness', and . . . opening up popular access to them (Bianchini, 1993, p. 19).

Flagship schemes, as he continues, enhance a city's competitiveness while grassroots culture requires a decentred approach, so that the former tends to be supported at the cost of the latter. Similarly, Gonzalez writes of Bilbao that its culturally led redevelopment relies on

> ephemeral spectacles, aimed at attracting and encouraging the development of local cultural industries [so that culture is] expressed in the language of economics and would serve economic development objectives (Gonzalez, 1993, p. 85).

And Jude Bloomfield writes of Bologna

> It has proved easier to solve the problems of the new middle-class youth by enabling them to become cultural entrepreneurs than that of bridging the gap between them and the poorly skilled and alienated underclass (Bloomfield, 1993, pp. 111–112).

Zukin argues that the power to create an image of a city has increased in importance when social classes and political parties

> have become less relevant mechanisms of expressing identity. Those who create

images stamp a collective identity. Whether they are media corporations like the Disney Company, art museums, or politicians, they are developing new spaces for public cultures (Zukin, 1995, p. 3).

Elsewhere, she concludes that

> Far from suggesting a free expression of divergent identities, the flourishing of new cultural meanings in the highly competitive environment of urban space makes it more urgent to understand their material effects (Zukin, 1996, p. 242).

From a similarly critical position, in the UK, artists' group Hewitt + Jordan—artists Andy Hewitt and Mel Jordan, based in Sheffield—write

> Cultural policy can be divisive. Culture-led regeneration is only representative of a wider constituency and wider culture of the city when it is developed alongside a social policy that stems from a vigorous and democratic political process. This demands a political system that has the confidence to take on and discuss the bigger and longer-term problems affecting the city (Hewitt + Jordan, 2004, p. 29).

I doubt this system is currently available in the UK or US. The result is that cultural production is co-opted by developers and governments alike to provide badges of respectability for practices which may produce social division rather than equity. The use of culture in culturally led urban development trades on culture's supposed universal value to render its commissioning beyond contest while redevelopment itself may be highly contestable, as in the construction of what Jon Bird (1993) has called 'Dystopia on the Thames' in London's docklands. Landry and Bianchini assert that "Seemingly superficial, 'cosmetic' interventions can have an important effect on morale" (Landry and Bianchini, 1995, p. 31), but is the effect sustainable? Urban regeneration implies a social base and may not be open to top–down or design solutions, or creativity takes resistant forms. Rose Gilmore notes a poster which people

are invited to copy and display in protest against increased penalties for flyposting, in the Rope Walks Quarter of Liverpool: "THIS IS CULTURE" (Gilmore, 2004, p. 128).

1.6 The Cultural Industries or the Culture Industry?

Writing in the 1940s, Teodor W. Adorno and Max Horkheimer (1947/1997) refused the term 'mass culture' because it was agreeable, as Adorno later recalls, to the proprietors of an industry of mass deception. It may seem odd that I introduce this, having criticised Kelly for reviving Lynch. A historical adjustment is required in reading Adorno. After cultural studies, from the Birmingham School in the 1960s onwards, his rejection of intermediate art forms such as film and jazz seems quaint. For cultural studies, any area of cultural production—comics to Racine—is useful in articulating the received or contested values of a period. This has never meant that a play by Racine is equal aesthetically to *graffiti*, but recognises the specifics of aesthetic criteria in context of their social production; *and allows that intervention in the production of categories is an intervention in the production of society.* Adorno's refusal of the term mass culture is an example of such intervention. His critique is conditioned also by the Nazis' closure of the Frankfurt Institute for Social Research and its reinstitution at Columbia University, New York; and if the spectre which haunts the Frankfurt School as a whole is the rise of fascism in industrialised Germany after the failure of the German Revolution in 1919, the prevailing condition which inflects Adorno's work in the 1940s is his exposure, as a European intellectual Jewish Marxist, to the movies, radio and the popular press in the US. Part of his response was a detailed interrogation of the horoscope column of *The Los Angeles Times* (Adorno, 1994, pp. 34–127) in which he writes that the advice given

> implies that all problems due to objective circumstances such as, above all, economic

difficulties, can be solved in terms of private individual behaviour or by psychological insight (Adorno, 1994, p. 57).

Perhaps this could inform a reconsideration of the cultural industries. This is not an argument against psychological, or more specifically psycho-analytic, investigations of urban cultural experiences (see Sibley, 2001, for such an approach), but draws attention to the non-deliberative ways in which culture is consumed. Whether consumers are dupes of the market or knowing manipulators of the manipulation it offers them is the subject-matter of a considerable literature in sociology (see Oh and Arditi, 2000, for a summary of positions). For Adorno it is the former; but for Walter Benjamin cinema offers audiences imaginative opportunities to remake the plot in awareness of the alienating labour of actors who make a film in multiple takes. But Benjamin did not go to the movies but to Paris, and to Moscow in the 1920s to see experimental theatre and film as vehicles of revolution. It is more that Adorno and Benjamin are writing about different things than that they disagree; Adorno's rejection of mass culture is not undermined by Benjamin's celebration of a democratic lens.

Benjamin's work is more often used in cultural studies teaching. His essay on 'The work of art in an age of technical reproducibility', to use a correct translation of the title (Benjamin, 1970, pp. 219–254; see Leslie, 2000, p. 132) is a standard text. Possibly more interesting is his address to a group of anti-fascist writers in Paris in 1934, published as 'The author as producer' (in Benjamin, 1983) in which he alludes to the contributions of readers as authors in the Soviet press, a case of mass culture in which the medium is reclaimed by its public. Again, his experience differs from Adorno's, listening uncomfortably to the reduction of classical music to easy listening on the wireless. It is to Adorno that I turn nonetheless because his refusal of the term 'mass culture', reiterated in a reconsideration of the 1947 essay in 1975 (Adorno, 1991, pp. 85–92), is key to my critique of today's cultural policy.

In approaching this material, nonetheless, I need to be selective. For example

The culture industry fuses the old and familiar in a new quality. In all its branches, products which are tailored for consumption by masses, and which to a great extent determine the nature of that consumption, are manufactured more or less according to plan ... The culture industry intentionally integrates its consumers from above. To the detriment of both it forces together the spheres of high and low art, separated for thousands of years (Adorno, 1991, p. 85).

This extract links a critique of the culture industry as dominating the consumption of cultural products with a concern that high and low art forms are fused. A limitation of Adorno's position is that high art cannot be produced *by*, only for, its public by specialist interpreters. The point I take, however, is that mass culture is mass deception and begins in the term mass culture as if it is produced by consumers while the hold over broadcasting technology by commercial radio stations in the US in the 1940s, Hollywood over the movies and companies such as Disney and Time-Warner over global media now, means that cultural consumption is determined not by listeners and viewers but by the owners of the means of production.

The market, not social need, drives industries such as film and fashion, and increasingly subsumes high art (see Leslie's comment on Tate Modern above). Adorno's argument seems valid when he writes that "The entire practice of the culture industry transfers the profit motive naked onto cultural forms" (Adorno, 1991, p. 86). It is not a surprise: the function of industry, cultural or otherwise, in capitalism is to increase wealth, not make people happy. Yet the culture industry gives an illusion of choice in adherence to a conformity determined by the market and—bearing in mind the discussion above—a by-product of the culture industry is an image of a conflict-free society which masks divergences of power and need the negotiation or contestation of which are legitimate aspects of public, political and *cultural* concern.

On the former, Adorno writes in 'The Schema of Mass Culture', published in German in 1981

> The dream industry does not so much fabricate the dreams of the customers as introduce the dreams of the suppliers among the people. This is the thousand-year empire of an industrial caste system governed by a stream of never ending dynasties (Adorno, 1991, p. 80).

There is a resonance: empire translates as *reich*. The passage recalls this from the 1940s

> The culture industry perpetually cheats its consumers of what it perpetually promises. The promissory note which ... it draws on pleasure is endlessly prolonged; the promise, which is actually all the spectacle consists of, is illusory: all it actually confirms is that the real point will never be reached, that the diner must be satisfied with the menu (Adorno and Horkheimer, 1947/1997, p. 139).

Adorno recognises the liberating bourgeois characteristic of intrigue in drama—a way to renegotiate the conditions of a subject's life—but remarks that when heroes no longer make sacrifices or come of age, but achieve a success which affirms conformity, the intriguer is liquidated. He writes that mass culture "treats conflict but in fact proceeds without conflict ... reality becomes a technique for suspending its development"; and that "Mass art registers this fact inasmuch as it repudiates conflict as outmoded" (Adorno, 1991, pp. 62, 66). I put this beside Zukin's argument that when

> labour unions and political parties seem powerless to challenge social divisions, culture as 'collective lifestyle' appears a meaningful, and often conflictual, source of representation [while culture is] often reduced to a set of marketable images (Zukin, 1995, p. 263).

And Ian Angus' comment that mass culture replaces a divided class-based culture "with a single self-enclosed world of industrially produced cultural goods" so that, as cultural uniformity drives out regional, ethnic and linguistic difference, inequalities are "expressed not as different worlds of goods, but as *relative degrees of access* to uniform goods" (Angus, 2000, pp. 89–90). I would cite, too, remarks by a dissident cultural professional in Glasgow

> The wish locally to bury the facts of a past which had become inconvenient and to superimpose a new, sanitised, marketable image of the city required not a critical social history ... but a bland, self-congratulatory hype (quoted in Bianchini, 1991, p. 37).

Why reintroduce critical theory in urban studies? It lacks an empirical base (and I have suggested arts advocacy suffers from this condition) and offers only a discursive, dialectical approach to material problems. An underpinning theme in Adorno's writing (1997) is the tension between art's aesthetic and social dimensions, (dis)coloured by a view that in dark times art, too, will be dark in as much as it critically conveys the absurdities of the administered world. For Adorno, the plays of Samuel Beckett stand for such a response. Yet, if Beckett is as acquired a taste as modern art, the concern is with consciousness (and conscience); it is also with dialectics—the insight that while people are shaped by conditions they also intervene in and reshape the conditions which shape them. Put more simply, Benjamin's idea of the author as producer, and citizen as co-author, can be adapted to express an idea of the dweller as co-producer of urban spaces. This is compatible with, if different from, Lefebvre's reflection on lived space. In brief, then, I would argue that the method of interrogation advanced by Adorno can be applied to culturally led urban redevelopment and that part of his critique retains validity in that context (of mass deception then and the illusions of cultural consumption and the commodification—as aestheticisation—of urban spaces now).

2. Art as Interruption

Despair is easy—the more so after reading Adorno—but observing artists' groups working against the grain of cultural universalism gives me hope. The work is seldom a resolution of a situation, more often it problematises one. Matthew Cornford and David Cross, as the London-based group Cornford and Cross, describe their aim as to

> incite individuals to collude in encroachments upon their own freedom in the urban environment, and so critically engage with the relationship of art to the exercise of power in public space (quoted in Neilson *et al.*, 2004, p. 26).

They do not seek a mass public but to engage specific publics for whom imaginative possibilities are opened.

To give an example of the kind of project I have in mind, Hewitt + Jordan produced a billboard text for a site at the corner of Corporation Street and Alma Street, Sheffield, in April, 2004 which states that "The economic function of public art is to increase the value of private property" (Hewitt + Jordan, 2004, p. 53; see Figure 1). The artists see the work as setting out "to question the function of art in the public realm within the economic regeneration of post industrial cities" (www.jordan-hewitt.demon.co.uk). It is the second part of a project which began with the artists presenting themselves to delegates at a conference in London in 2003 as the prize in a raffle. Both works were commissioned by Public Art Forum, a network of agencies, public authorities and individuals involved in public art's management. Hewitt + Jordan do not make documentary objects out of such projects—in the way Andy Goldsworthy, say, relies on gallery-shown colour photographs of works in remote places to bring them to a public and make a living as an artist—and state

Figure 1. The economic function of public art is to increase the value of private property. By courtesy of the artists Hewitt and Jordan.

We always try to avoid making something—not even a video—not that we can make video anyway. We know that we are making it difficult for ourselves. I think that the reason for this is a desire to focus the attention on the intervention/process itself rather than on an object—an object brings 'relief' to the normal spectator of art (Hewitt + Jordan, 2004, p. 47).

In a more recent project, *Futurology*, Hewitt + Jordan commissioned a range of other artists to work in partnership with schools in the West Midlands and arranged a series of critical discussions of culture at the New Art Gallery in Walsall. The project was supported by Creative Partnerships, but Jordan and Hewitt saw it as, in itself, a critique of the intention for which Creative Partnerships was established by government: to "encourage creative learning in schools" (Futurology, 2004). Rather than ask what is uncreative learning I quote the project leaflet

Current government believes this form of learning will help a future citizen adapt to the changing economic environment. Creative Partnerships is aimed at bringing this learning initiative to schools in areas that are the most 'economically and socially challenged' … We want to avoid both the cynical withdrawal of artists from the public as well as the naïve surrendering of the artist to the agenda of politicians and funders (Futurology, 2004).

That agenda tends to employ artists as low-budget problem-solvers, sometimes putting them in situations in which they have little chance to contribute—through short-term and peripatetic involvement—to structural problems which may in any case result from other government policies. In a conversation, Hewitt and Jordan mentioned that a by-product of Creative Partnerships and other socially directed schemes is a worry on the part of government arts officers that there is a lack of evidence that quantifiable benefits accrue (conversation with the author, 11 August 2004, Walsall). This seems to be a rerun of the problem identified by Selwood

(1995) and could offer a reclamation of the autonomy characteristic of Modernism in the arts, or imply a deeper questioning of the conditions in which social dis-ease arises. Hewitt + Jordan look to a reformed autonomy

The fact that our practice attempts to be about some idea of art's transformation … is also crucial to this idea of the autonomy of art. But I don't mean that in the sense of wanting it to be outside everyone's experience … There are good things about being autonomous too—like an objective view and a dissident voice (Hewitt + Jordan, 2004, p. 44).

Jordan, speaking there in an interview, does not adopt an activist position but one of seeking to change art rather than the world, as a possible means to change part of the way in which the idea of a world (distinct from the bio-realm of Earth) is constructed. Patricia Phillips, from a different viewpoint, writes of new modes of practice in north America which have

produced a variety of social, political, and activist forms—installations, interventions, roundtables, performances, and multiple forms of collaboration that engage urgent subjects (housing and homelessness, social justice, domestic violence, race and class, forgotten histories and untold stories) in a passionate, of eclectic hybridity (Phillips, 2003, p. 12).

In the UK, too, a range of eclectic and hybrid practices has begun to emerge. The work of the London-based group PLATFORM (Jane Trowell, James Marriott, Dan Gretton and Emma Sangster) links art to democracy and ecology. The group have as many links to campaigning organisations as to the artworld and refuse to follow the agendas of arts funding bodies regardless of what has at times been a fluctuating financial position. Their current preoccupation is with the global impact on human and environmental rights of the oil industry and projects have included production and distribution to commuters of a spoof newspaper and guided walks around the financial district of London (see Miles, 2004a, pp. 195–203).

If PLATFORM politicises the fantasy of an ever-expanding global economy which is materially destructive of human rights and natural habitats on a correspondingly global scale, there is also work which interrupts the rhetoric of urban development in a localised way—such as *Camelot*, a project by Cornford and Cross, comprising an industrially made steel security fence erected around a residual patch of grass near the bus station in Stoke-on-Trent in 1996. During its installation, the project caused huge resentment of its form, which obstructed informal paths to the shopping centre, and of the use of public money to pay for it. It could have been (and was for some) a public relations disaster. Yet it led to intensive discussions on the quality and ownership of public spaces in the city and their neglect by the local authority, discussions which were perhaps elements of a more direct democracy than that of local elections. The artists state

> *Camelot* is a literal interpretation of the City Limits theme; we chose to invite reflection and debate on the physical and social boundaries which often determine the patterns of city life—in this case by denying people access to some small, neglected fragments of public urban land (artists' statement; in Miles, 2004a, p. 166).

This makes any audit of the work as might be required on conventional public art problematic and denies the kind of solution-based evaluation which was required in, for instance, SRB-funded projects and is now required for Creative Partnerships. Indeed, I have only the word of one of the artists involved as to what took place.

An implication, tacit or stated, of such work is that it contributes to conditions in which radical socioeconomic change is possible. There are at least two ways in which to interpret this: either the change is personal—as the personal is political—and occurs in more or less intimate exchanges between artists and micro-scale, participating audiences, in which context PLATFORM arrange events on the parallel worlds of the oil industry and the management of the Holocaust for groups of around eight; or intervention is in the categories and conventions of discourse in order to shift how specific issues are represented or (re)considered. The former might rely on a latent utopianism which is part of modern culture's heritage, but this needs scrutiny. Above, I cite Hope and Carrington, the group B + B, on the reliance of the integration of art into the current UK government's policy on social inclusion and regeneration on such utopian notions of art as empowerment (www.welcomebb.org.uk). They also note that Joseph Beuys reframed art therapeutically to empower people to live creatively, which I take to include imagining futures other than those prescribed by capital or its out-sourced providers of governmental services in a globalised economy. But what Beuys meant by creativity is not what is meant in government policy, in what Hope and Carrington call the art pill. B + B further cite Alan Kaprow, who built 30 ice walls in Pasadena and Los Angeles in 1967 as dystopian spectacles resembling capitalism.

The approaches sketched here denote a stratum of cultural production which crosses the boundaries of art and social formation, and becomes a form, too, of cultural mediation when artists take responsibility for the dissemination of their own projects. It probably no longer matters whether such activity is classified as art, except that artists need arts funding (or jobs in arts education) for support even when they set out to critique such support systems. A future project for Hewitt + Jordan is to distribute 422 300 badges—one for everyone—in Manchester saying "I will not accept 'the way things are'" (badge in author's possession). Is it cool or should I note Arundhati Roy's complaint that, as a writer taking sides over the construction of highly destructive dams in India, "that's considered a pretty uncool, unsophisticated thing to do ... uncomfortably close to the territory occupied by political party ideologues" (Roy, 2001, p. 11)? Does cultural engagement correspond to the agenda of the World Commission on Culture and Development, for whom cultures are "ways of living together" (UNESCO, 1996, p. 14), more than to that of developers and cultural institutions? Perhaps we can think of cultural work in which production and reception link in participatory ways of working as development in the sense of development in the non-affluent world, where NGOs have sometimes been able to take more empowering approaches than those of urban policy in the affluent world (Guha and Martinez-Alier, 1997, for example). But that is another enquiry and might begin by looking at culturally led redevelopment via the literatures of sustainability and development studies. I note a summary of the World Commission's report

> One of the most basic freedoms is to be able to define our own basic needs. This freedom is threatened by a combination of global pressures and global neglect ... Awareness of this has led to resurgent assertions in the post-Cold War world, as people and their leaders turn to their own culture as a means of self-definition and mobilisation. For the poorest among them, their own values are often the only thing that they can assert ... The concern is that development has meant loss of identity, sense of community and personal meaning (UNESCO, 1996, p. 15).

There are traces of cultural liberalism in UNESCO, yet the report begins to separate culture as way of life, but also as action, from the demands of a globalised economy and the culture it produces, and requires, in order to glue that economy together and keep it going. Culture, after all, is arguably more influential in establishing brand loyalties by turning products into iconic representations of an alluring lifestyle than simple advertising (although that is one channel by which the representation is enforced, of course). Looking to the literatures of development in the non-affluent world, where the term 'development' is brought into contestation in debates on sustainability, more radical approaches are found than in most of the reports, schemes and image-constructions of urban

redevelopment in the affluent world. The concept of liberation ecology, to take a particularly interesting case, is a discourse of Nature which is "Marxist in origin, poststructural in recent influence, politically transformative in intent, but subject still to the fiercest of debates" (Peet and Watts, 1996, p. 37). I wonder if the concept can be mapped onto the affluent world to subvert its notions of development, whether culturally or economically led. In the non-affluent world, the affluent world's notion of development as mono-crop agriculture is increasingly rejected by aid agencies for whom handing over management to local groups is imperative if solutions to environmental degradation are to be lasting. Summarising the position, Elliott writes

> Ensuring that individual land users and communities have secure rights to resources and the benefits from investments therein is a further condition of sustainable agricultural development based on recent experiences of success (Elliott, 1999, p. 126).

Could that be applied to deprivation in an inner-city area? It would perhaps not produce the solution advanced naively in 1995 by Landry and Bianchini for the revitalisation of urban centres

> In Newcastle they have used the Happy Hour ... On various days of the week, some bars and restaurants reduce their prices substantially in order to encourage people to stay in the city centre and use its facilities. The prices are so low that it is hardly worth going home and cooking your own food (Landry and Bianchini, 1995, p. 42).

In Newcastle, too, and in several other city centres in the UK, police chiefs now seek additional powers to curb the effects of excess alcohol consumption among young people attracted there by, precisely, happy hours. The dream of oblivion may be more easy than most to market and has the advantage of perpetual non-satisfaction in consequent amnesia, but is a degradation of urban dwelling equivalent, for the same profit motive, to, say, logging in Indonesia. My con-clusion—it may be inept to venture a con-clusion in discussion of a process which I have said from complexity theory is unpre-dictable—is that the values of the contempor-ary art cited above are more akin to those of the work of some NGOs in the non-affluent world, or of activists resisting globalised capital and its environmental destructiveness and human rights abuses, than to those of urban redevelopment in the affluent world or modern art.

A key component in this is the handing over to participants of co-production of the work. A similar departure is found in radical planning when the planner retains her/his expertise but relinquishes the safety of both statistics and the office to spend time with mobilised community groups. Sandercock writes that

> Radical practices emerge from experience with and a critique of existing unequal relations and distributions of power, oppor-tunity, and resources (Sandercock, 1988, p. 97).

It is important to note that radical planners and activist artists do not cease to be planners and artists but do accord equal value to both the expertise of professionals in their fields *and* dwellers on dwelling. These practices interrupt the flow of city-image rhetoric. Moments of close conversation on the archi-tectures of power, say, during a guided walk through London's financial district are not given to city marketing, any more than a city which is promoted as having no grand design to articulate its narrative will attract mainstream investment. Neither are the activi-ties of Cornford and Cross likely to put Stoke on the international culture or tourism maps. Their provocation of reconsiderations of the values of the public realm contrast with fanta-sies of a latte-drinking, piazza-sitting society. It could be compared with the work of the Yes Men, Andy Bichlbaum and Mike Bonano, ex-media teachers, who construct an ironic cri-tique of global trade by impersonating it via a spoof World Trade Organisation website—from which they are invited to address

business gatherings. A press report notes that, impersonating (or identity correcting, as they put it) a representative of McDonalds for a student audience, Bonano

> strolls over to his overhead projector and begins to outline his firm's latest act of cor-porate responsibility: converting first-world human waste into fast food for developing countries (Burkeman, 2004).

After two decades of public art, the inte-gration of artists in the design of highways and bridges, their complicity in public–private finance initiatives and the view that culture solves socioeconomic problems which may result from other areas of public policy, I find that refreshing.

But I need to ask one more question: is the approach I cite above another form of cultural expediency differing only in specifics from its predecessors, just as much avant-garde art since the 19th century reproduced the conven-tions (such as the privileged insight of the artist) of the social arrangements it sought to overthrow? Yúdice writes

> In our era, representations of and claims to cultural difference are expedient insofar as they multiply commodities and empower community (Yúdice, 2003, p. 25).

Is radical culture, then, a resource? As such, it must let go of the Modernist claim to auton-omy even within radical contemporary art practice. At the same time, it retains the equally Modernist claim to deal in privileged insights. Yúdice traces, from Foucault, an evolution of a relation between thought and the world (*episteme*) in which the post-Enlightenment phase is characterised by het-erodox enquiry and a redemption of "great hidden forces" (Foucault, 1973, p. 251, quoted in Yúdice, 2003, p. 30). Yúdice continues

> Modern knowledge thus consists of unveil-ing the primary processes (the infrastruc-ture, the unconscious) that lurk in the depths, beneath the surface: manifestations of ideology, personality, and the social (Yúdice, 2003, p. 30).

and proposes a next phase in which previous modes (resemblance, representation and historicity) are recombined to account for "the constitutive force of signs". The rupture of society predicted in Marxist analyses of capitalism is salved by cultural expedients and "the transformation of artists and intellectuals into managers of that expropriation under the guise of 'community-based' work" (Yúdice, 2003, p. 35). The prognosis is as gloomy as Adorno's

> That culture as resource is at the heart of these processes does not mean that capital's assault on workers and others ... [is] merely virtual. It is for this reason that cultural politics ... is unlikely to make a difference. Indeed, I argue ... that the 'cultural left' is largely enjoined to perform such a cultural politics ... The protection of the cultural resources that global entertainment conglomerates have expropriated involves not only the law but also the use of police and military forces, for example, in the pursuit of piracy ... From the perspective of most forms of cultural politics ... subversion of the assumptions implicit in dominant media as a way of appropriating them is thought to be a viable option ... [but] it is hardly effective (Yúdice, 2003, pp. 35–36).

His inclination, as I read it, is that music piracy is a more frontal assault on capitalism than the kinds of intervention I have sketched above. Interestingly, music piracy is both an extension and a counter to the entrepreneurial activities of the cultural intermediaries O'Connor sees as vitalising new urban spaces.

In the end, I admit I cling to hope because a world without it is too awful to contemplate, not from evidence. But then Zukin reports that

> the belief that New York is the world capital of culture has been used as if it were a fortune-teller's benediction to ward off all evidence of economic decline (Zukin, 1995, p. 110).

Nonetheless, I look to art which takes a dialectical approach as a viable alternative to either complicity in or frontal resistance to globalised capitalism and its cultural turn. For Jane Trowell this work has "a viral quality, slipping a proposition into the bloodstream under the guise of a safe publication" (Trowell, 2000, p. 107). Perhaps the spoof newspaper and the spoof website which mimic in order to refuse the appearances of an increasingly total neo-liberalism, appearances which today have the function of the deceptions of the culture industry previously, are interventions in the texture of globalised communications, interrupting its gloss and acting, almost imperceptibly, like frost in concrete. Perhaps that is what is viable now.

References

ADORNO, T. W. (1991) *The Culture Industry: Selected Essays on Mass Culture*, Ed. by J. M. Bernstein. London: Routledge.

ADORNO, T. W. (1994) *The Stars Down to Earth and Other Essays on the Irrational in Culture*, Ed. by S. Crook. London: Routledge.

ADORNO, T. W. (1997) *Aesthetic Theory*, trans. by H. Hullot-Kentor. London: Athlone.

ADORNO, T. W. and HORKHEIMER, M. (1947/1997) *Dialectic of Enlightenment*. London: Verso.

ANGUS, I. (2000) *Primal Scenes of Communication: Communication, Consumerism, and Social Movements*. Albany, NY: State University of New York Press.

ARTS COUNCIL (1989) *An Urban Renaissance*. London: Arts Council.

BELL, D. and JAYNE, M. (2004) *City of Quarters: Urban Villages in the Contemporary City*. Aldershot: Ashgate.

BENJAMIN, W. (1970) *Illuminations*, Ed. by H. Arendt. London: Fontana.

BENJAMIN, W. (1983) *Understanding Brecht*. London: Verso.

BENJAMIN, W. (1999) *The Arcades Project*, Ed. by H. Eiland and K. McLaughlin. Cambridge, MA: Harvard University Press.

BENNETT, S. and BUTLER, J. (Eds) (2000) *Locality, Regeneration & Diversities*. Bristol: Intellect Books.

BIANCHINI, F. (1991) Alternative cities, *Marxism Today*, June, pp. 36–38.

BIANCHINI, F. (1993) Remaking European cities: the role of cultural policies, in: F. BIANCHINI and M. PARKINSON (Eds) *Cultural Policy and Urban Regeneration: The West European Experience*, pp. 1–21. Manchester: Manchester University Press.

BIANCHINI, F. and PARKINSON, M. (Eds) (1993) *Cultural Policy and Urban Regeneration: The*

West European Experience. Manchester: Manchester University Press.

BIANCHINI, F., FISHER, M., MONTGOMERY, J. and WORPOLE, K. (1988) *City Centres, City Cultures*. Manchester: Centre for Local Economic Strategies.

BIRD, J. (1993) Dystopia on the Thames, in: J. BIRD, B. CURTIS, T. PUTNAM *ET AL.* (Eds) *Mapping the Futures*, pp. 120–135. London: Routledge.

BIRD, J., CURTIS, B., PUTNAM, T. *ET AL.* (Eds) (1993) *Mapping the Futures*. London: Routledge.

BLOOMFIELD, J. (1993) Bologna: a laboratory for cultural enterprise, in: F. BIANCHINI and M. PARKINSON (Eds) *Cultural Policy and Urban Regeneration: The West European Experience*, pp. 90–113. Manchester: Manchester University Press.

BOOTH, P. and BOYLE, R. (1993) See Glasgow, see Culture, in: F. BIANCHINI and M. PARKINSON (Eds) *Cultural Policy and Urban Regeneration: The West European Experience*, pp. 21–47. Manchester: Manchester University Press.

BORDEN, I., KERR, J., PIVARO, A. and RENDELL, J. (Eds) (1996) *Strangely Familiar: Narratives of Architecture in the City*. London: Routledge.

BORRIES, F. von (2003) Consumption and the post-industrial city: Nike Town, in: M. MILES and N. KIRKHAM (Eds) *Cultures & Settlements*, pp. 75–86. Bristol: Intellect Books.

BRISTOL LEGIBLE CITY (2001a) *From Here to There*. Bristol: Bristol Legible City.

BRISTOL LEGIBLE CITY (2001b) *You Are Here*. Bristol: Bristol Legible City.

BURKEMAN, O. (2004) The Bush baiters, *The Guardian*, review section, 2 November, p. 8.

BYRNE, D. (1997) Chaotic places or complex places? Cities in a postindustrial era, in: S. WESTWOOD and J. WILLIAMS (Eds) *Imagining Cities: Scripts, Signs, Memories*, pp. 50–72. London: Routledge.

CILLIERS, P. (1998) *Complexity and Postmodernism: Understanding Complex Systems*. London: Routledge.

CLINE, A. (1997) *A Hut of One's Own*. Cambridge, MA: MIT Press.

DEGEN, M. (2002) Regenerating public life? A sensory analysis of regenerated public places in El Raval, Barcelona, in: J. RUGG and D. HINCHCLIFFE (Eds) *Recoveries and Reclamations*, pp. 19–36. Bristol: Intellect Books.

DEGEN, M. (2004) Barcelona's Games: the Olympics, urban design, and global tourism, in: M. SHELLER and J. URRY (Eds) *Tourism Mobilities: Places to Play, Places in Play*, pp. 131–142. London: Routledge.

DEUTSCHE, R. (1996) *Evictions: Art and Spatial Politics*. Cambridge, MA: MIT Press.

DODD, D. (1999) Barcelona, the making of a cultural city, in: D. DODD and A. VAN HEMEL (Eds) *Planning Cultural Tourism in Europe: A Presentation of Theories and Cases*, pp. 53–64. Amsterdam: Boekman Stichting.

DODD, D. and HEMEL, A. VAN (Eds) (1999) *Planning Cultural Tourism in Europe: A Presentation of Theories and Cases*. Amsterdam: Boekman Stichting.

ELLIOTT, J. (1999) *An Introduction to Sustainable Development*, 2nd edn. London: Routledge.

EVANS, G. (2001) *Cultural Planning: An Urban Renaissance?* London: Routledge.

EVANS, G. (2004) Cultural industry quarters: from pre-industrial to post-industrial production, in: D. BELL and M. JAYNE (Eds) *City of Quarters: Urban Villages in the Contemporary City*, pp. 71–92. Aldershot: Ashgate.

FOUCAULT, M. (1973) *The Order of Things: An Archaeology of the Human Sciences*. New York: Vintage.

FUTUROLOGY (2004) Project leaflet produced by Hewitt + Jordan, The New Art Gallery, Walsall.

GALLAGHER, A., PHILLIPS, A. and RENTON, A. (Eds) (2004) *Tales of the City*. Bologna: Arte Fiera 2004.

GILMORE, A. (2004) Popular music, urban regeneration and cultural quarters: the case of the Rope Walks Quarter, Liverpool, in: D. BELL and M. JAYNE (Eds) *City of Quarters: Urban Villages in the Contemporary City*, pp. 109–130. Aldershot: Ashgate.

GOLDMAN, R. and PAPSON, S. (1998) *Nike Culture*. London: Sage.

GONZALEZ, J. M. (1993) Bilbao: culture, citizenship and quality of life, in: F. BIANCHINI and M. PARKINSON (Eds) *Cultural Policy and Urban Regeneration: The West European Experience*, pp. 73–89. Manchester: Manchester University Press.

GOTTDIENER, M. (Ed.) (2000) *New Forms of Consumption: Consumers, Culture, and Commodification*. Lanham, MD: Rowman & Littlefield.

GUHA, R. and MARTINEZ-ALIER, J. (1997) *Varieties of Environmentalism: Essays North and South*. London: Earthscan.

HALL, T. and HUBBARD, P. (Eds) (1998) *The Entrepreneurial City: Geographies of Politics, Regime and Representation*. Chichester: Wiley.

HARRIS, S. and BERKE, D. (Eds) (1997) *Architecture of the Everyday*. New York: Princeton Architectural Press.

HEWITT + JORDAN (2004) *I Fail to Agree*. Sheffield: Site Gallery.

HILL, J. (Ed.) (1998) *Occupying Architecture*. London: Routledge.

JACOBS, J. (1961) *The Death and Life of Great American Cities*. New York: Random House.

KELLY, A. (2001) *Building Legible Cities.* Bristol: Bristol Legible City.

LANDRY, C. (2000) *The Creative City: A Toolkit for Urban Innovators.* London: Earthscan.

LANDRY, C. and BIANCHINI, F. (1995) *The Creative City.* London: Demos.

LARSON, G. O. (1997) *American Canvas.* Washington, DC: National Endowment for the Arts.

LEFEBVRE, H. (1991) *The Production of Space.* Oxford: Blackwell.

LESLIE, E. (2000) *Walter Benjamin: Overpowering Conformism.* London: Pluto.

LESLIE, E. (2001) Tate Modern: a year of sweet success, *Radical Philosophy,* 109, pp. 2–5.

LOFTMAN, P. and NEVIN, B. (1998) Pro-growth local economic development strategies: civic promotion and local needs in Britain's second city, 1981–1996, in: T. HALL and P. HUBBARD (Eds) *The Entrepreneurial City: Geographies of Politics, Regime and Representation,* pp. 129–148. Chichester: Wiley.

LYNCH, K. (1960) *The Image of the City.* Cambridge, MA: MIT Press.

MARCUSE, H. (1937/1968) The affirmative character of culture, in: *Negations,* pp. 88–133. Harmondsworth: Penguin.

MASSEY, D. (1994) *Space, Place and Gender.* Cambridge: Polity.

MERRIFIELD, A. and SWYNGEDOUW, E. (Eds) (1996) *The Urbanization of Injustice.* London: Lawrence & Wishart.

MILES, M. (2000) *The Uses of Decoration: Essays in the Architectural Everyday.* Chichester: Wiley.

MILES, M. (2004a) *Urban Avant-gardes: Art, Architecture and Change.* London: Routledge.

MILES, M. (2004b) Drawn and quartered: El Raval and the Haussmannization of Barcelona, in: D. BELL and M. JAYNE (Eds) *City of Quarters: Urban Villages in the Contemporary City,* pp. 37–55. Aldershot: Ashgate.

MILES, M. and KIRKHAM, N. (Eds) (2003) *Cultures & Settlements.* Bristol: Intellect Books.

MYERSCOUGH, J. (1988) *The Economic Importance of the Arts in Britain.* London: Policy Studies Institute.

NEILSON, E., DAVIES. L.-R. and OFFERS, S. VON (2004) Cornford and Cross, in: A. GALLAGHER, A. PHILLIPS and A. RENTON (Eds) *Tales of the City,* pp. 26–27. Bologna: Arte Fiera.

O'CONNOR, J. and WYNNE, D. (Eds) (1996) *From the Margins to the Centre: Cultural Production and Consumption in the Post-Industrial City.* Aldershot: Ashgate.

O'CONNOR, J. (1998) Popular culture, cultural intermediaries and urban regeneration, in: T. HALL and P. HUBBARD (Eds) *The Entrepreneurial City,* pp. 225–240. Chichester: Wiley.

OH, M. and ARDITI, J. (2000) Shopping and postmodernism: consumption, production, identity, and the Internet, in: M. GOTTDIENER (Ed.) *New Forms of Consumption: Consumers, Culture, and Commodification,* pp. 71–92. Lanham, MD: Rowman & Littlefield.

PEET, R. and WATTS, M. (Eds) (1996) *Liberation Ecologies: Environment, Development, Social Movements.* London: Routledge.

PHILLIPS, P. (2003) Unsettled sites: suspended attention. How is the city an issue for art?, in: L. PALMER (Ed.) *3 Acres: The Lake Du Sable Park Proposal Project,* pp. 12–15. Chicago, IL: White Walls.

ROY, A. (2001) *Power Politics.* Cambridge, MA: South End Press.

RUGG, J. and HINCHCLIFFE, D. (Eds) (2002) *Recoveries and Reclamations.* Bristol: Intellect Books.

SANDERCOCK, L. (1998) *Towards Cosmopolis.* Chichester: Wiley.

SCOTT, A. J. (2000) *The Cultural Economy of Cities.* London: Sage.

SELWOOD, S. (Ed.) (1995) *The Benefits of Public Art.* London: Policy Studies Institute.

SELWOOD, S. (Ed.) (2001) *The UK Cultural Sector: Profile and Policy Issues.* London: Policy Studies Institute.

SENNETT, R. (1998) *The Corrosion of Character: The Personal Consequences of Work in the New Capitalism.* New York: Norton.

SHAW, P. (1991) *Percent for Art: A Review.* London: Arts Council of Great Britain.

SIBLEY, D. (2001) The binary city, *Urban Studies,* 38(2), pp. 239–259.

SMIERS, J. (2003) *Arts Under Pressure: Promoting Cultural Diversity in the Age of Globalization.* London: Zed Books.

TAYLOR, B. (1993) From penitentiary to temple of art: early metaphors of improvement at the Millbank Tate, in: M. POINTON (Ed.) *Art Apart: Art Institutions and Ideology across England and North America,* pp. 9–32. Manchester: Manchester University Press.

TROWELL, J. (2000) The snowflake in hell and the baked alaska: improbability, intimacy and change in the public realm, in: S. BENNETT and J. BUTLER (Eds) *Locality, Regeneration & Diversities,* pp. 99–109. Bristol: Intellect Books.

UNESCO (1996) *Our Creative Diversity: Report of the World Commission on Culture and Development* (summary version). Paris: UNESCO.

UNIVERSAL FORUM OF CULTURES 2004 (nd) Project brochure. Barcelona: Universal Forum of Cultures—Barcelona 2004.

WESTWOOD, S. and WILLIAMS, J. (Eds) (1997) *Imagining Cities: Scripts, Signs, Memories.* London: Routledge.

WIGGLESWORTH, S. and TILL, J. (1998) The everyday and architecture, *Architectural Design*, profile 134, July/August.

WILLETT, J. (1967) *Art in a City.* London: Methuen.

WILLIAMS, R. (1973) *The Country and the City.* London: Chatto and Windus.

WILLIAMS, R. (1976) *Keywords: A Vocabulary of Culture and Society.* Harmondsworth: Penguin.

YÚDICE, G. (2003) *The Expediency of Culture: Uses of Culture in the Global Era.* Durham, NC: Duke University Press.

ZUKIN, S. (1995) *The Cultures of Cities.* Oxford: Blackwell.

ZUKIN, S. (1996) Cultural strategies of economic development and the hegemony of vision, in: A. MERRIFIELD and E. SWYNGEDOUW (Eds) *The Urbanization of Injustice*, pp. 223–243. London: Lawrence & Wishart.

'Our Tyne': Iconic Regeneration and the Revitalisation of Identity in NewcastleGateshead

Steven Miles

The de-industrialisation of cities has created a set of circumstances in which policy-makers throughout Europe and beyond have desperately sought to explore the possibilities for a post-industrial future. For many such cities, cultural investment in capital-intensive projects which make radical statements about where a city's future might lay, offer a promised land, but one that is ultimately often unrealisable. The development of Newcastle-Gateshead offers an example of an iconic culture-led project that appears, at least on the surface, to be succeeding. But can investment in iconic projects deliver what policy-makers ask of them? More pointedly perhaps, at what level, if at all, do such projects engage with the identity of a city and its people?

This article will address the impact of flagship regeneration projects and their role in radically rearticulating the meaning of place and space in a so-called post-industrial world. As Hunt points out

> The architectural critic Jonathon Glancey suggested that Victorian cities had created an urban culture on the back of their trade and industry, but today it is the other way around. Instead of culture springing from the inner workings of our cities, we see it as the way to make our cities work (Hunt, 2004, p. 350).

In what follows, it will be suggested that the success of investment in iconic cultural projects depends above all upon people's sense of belonging in a place and the degree to which culture-led regeneration can engage with that sense of belonging, whilst balancing achievements of the past with ambitions for the future.

Context

The impact of iconic developments on the re-emergence of deindustrialised communities is a matter of continued policy debate. John Prescott, Deputy Prime Minister has suggested that

> There's a quiet revolution taking place in our leading cities. Places that were once the engine room of the industrial revolution, employing millions in mills, factories, ports and shipyards, are learning new ways to create wealth in a global economy where brain has replaced brawn (DCMS, 2004, p. 12).

But there is undoubtedly a danger in assuming that cultural investment can provide some kind of an alternative future for all deindustrialised cities. This reflects a broader debate in which commentators such as Richard Florida have suggested that creativity has an increasingly significant role to play in the social and economic development of our cities and that

> regional economic growth is driven by the location choices of creative people—the holders of creative capital—who prefer places that are diverse, tolerant and open to new ideas (Florida, 2002, p. 223).

From this point of view, quality of place has overtaken quality of life as the factor in determining why creative people live where they live. The suggestion might therefore be that iconic projects provide tangible evidence of the quality of place. They are, in effect, symbols of a place in which creative people can feel they will belong. This certainly appears to be the feeling surrounding Liverpool, recently awarded 'Capital of Culture 2008' with Egbert Kossak, one of Europe's leading regeneration experts, commenting that

> The Fourth Grace will do for Liverpool what the Opera House has done for Sydney. Liverpool has won praise from around the world for preserving its historical buildings. The time is right for a new, iconic building which will represent the future (Liverpool, 2004, p. 1).

Regardless of the demise of the Fourth Grace, the optimistic tone here is a telling reflection of how those involved in the production of iconic cultural developments tend to perceive such projects. But it remains unclear how far the realities of urban regeneration can match up to the expectations both the Arts fraternity and policy-makers have of cultural investment on this kind of scale.

In recent years, the world's waterfronts have provided a particular focus for culture-led regeneration. Marshall (2001, p. 3) describes the waterfront as space "in the city which allows expressions of hope for urban vitality". He goes on to point out that in cities such as London, New York, Vancouver, Sydney and San Francisco waterfronts have historically been the staging-points for the import and export of goods, but that this is no longer the case in our information-saturated, service-oriented economies

> These waterfront redevelopment projects speak to our future, and to our past. They speak to a past based in industrial production, to a time of tremendous growth and expansion, to social and economic structures that no longer exist.... Through historical circumstance, these sites are immediately adjacent to centers of older cities, and typically are separated from the physical, cultural and physiological connections that exist in every city. They speak to a future by providing opportunities for cities to reconnect with the water's edge (Marshall, 2001, p. 5).

Much of the debate around the significance of iconic projects of this kind are tied up with concerns as to whether or not such investment can effectively ameliorate the consequences of deindustrialisation. In this context, McGuigan (1996) identifies a series of urban regeneration schemes frequently led by flagship cultural projects during the 1980s in cities such as Baltimore and later, in the UK, Liverpool, Manchester, Bristol and Cardiff (see Cowell and Thomas, 2002; Bassett et al., 2002). The problem with these sorts of developments, according to McGuigan is that they actually articulate the interests and tastes of the postmodern professional and managerial class without solving the problems of a diminishing production base, growing disparities of wealth and opportunity, and the multiple forms of social exclusion (McGuigan, 1996, p. 99).

Sharon Zukin (1991), meanwhile, refers to 'quixotic' urban renewal projects that simply remain unproven as far as their economic benefits might be concerned. Miles and Miles point out that new cultural institutions such as Tate Modern, the Guggenheim in Bilbao and Barcelona's Museum of Contemporary Art play a prime role as facilitators of cultural display, but perhaps more problematically, as signs of urban affluence

> Flagship cultural institutions, frequently financed as public sector investments to attract private-sector renovation of the surrounding area, tend to be engines not of democratisation of culture but of gentrification ... This is not all bad, in that run-down areas can be transformed, but it may displace a residual population unless it is adequately protected, and establishes a connection ... between cultural space and wealth accumulation (Miles and Miles, 2004, p. 53).

The social impacts of culture-led regeneration are not necessarily always positive. Even in those circumstances where positive impacts are assumed, causality is always uncertain. In this context, Vegara (2001) refers to the "miracle of Bilbao" in a necessarily tentative fashion. Back in 2001 the industrial decline of Bilbao was undoubted, but Vegara could only at this time go far enough to predict confidently that the conditions were such that Bilbao "could" arise form the ruins of its industrial past, not least as a result of the impact of the Guggenheim Museum. The broader sociological impact of cultural investment on this scale remains intangible and new lessons are constantly having to be learned. For example, in his discussion of the regeneration of Porto in the aftermath of its stewardship of the 'European Capital of Culture

2001' Balsas (2004) concludes that too much emphasis was put on attracting public investment to regenerate public space, replacing infrastructures and modernising cultural facilities, but at the expense of more fundamental institutional capacity building and civic creativity.

Communities of Culture?

The development of cities such as Bilbao and Porto represents both a localisation of global and economic social forces and a location in a world capitalist order as Zukin (1991) points out. The success of such developments is perhaps dependent upon the degree to which the reinvention of the urban landscape fits in with, rather than being foisted upon, the identity of the place concerned. For this reason, the notion of community is crucial. Authors such as Harvey (1990) have described how the post-modern condition has led to the 'end of community', while Delanty (2003) highlights the role of the global city in displacing urban communities. It could be argued that in a global age cultural investment can at least potentially provide a means of revitalising communities by providing them with a new so-called post-industrial future that can help them readjust to the new economic conditions in which they find themselves. As Delanty suggests

> Community is communicative in the sense of being formed in collective action based on place, and is not merely an expression of an underlying cultural identity (Delanty, 2003, p. 71).

From this point of view, local identities are socially constructed rather than just being identified with a locality simply because it happens to be there.

But culture-led regeneration will not automatically engage with local communities. An alternative interpretation would indeed be that a lot of culture-led investment inevitably produces placeless forms of cultural representation (Dicks, 2003). From this perspective, culture-led regeneration projects all too often rely on formulaic development plans producing standardised results; what Short (1989) calls the new international blandscape "sterile and lacking in imagination" (Owen, 1993, p. 15). Such cities are only distinguished from each other on artificial grounds—grounds constructed symbolically by the marketeer

> The ready-made identities assigned by city boosters and disseminated through the mass media often reduce several different visions of local culture into a single vision that reflects the aspirations of a powerful elite and the values, lifestyles, and expectations of potential investors and tourists. These practices are thus highly elitist and exclusionary, and often signify to more disadvantaged segments of the population that they have no place in this revitalized and gentrified urban spectacle (Broudehoux, 2004, p. 26).

Dicks (2003, p. 82) points out that the underlying rationale behind flagship redevelopmet projects is, in the above context, to generate new consumer demand by attracting new visitors and shoppers to the city and thus "is rarely directed primarily at improving the quality of life of existing residents". Dicks discusses the redevelopment of Cardiff Bay as an example of regeneration that could be accused of distancing the project from its locality and thus from the existing local culture. Meanwhile, Broudehoux goes on to argue that a city's cultural capital cannot be easily manipulated, insofar as inappropriately blatant image construction will inevitably give rise to tensions and political conflicts. In effect, the representation of a city must do more than simply construct a 'pseudo-place' (Augé, 1995).

It will be suggested in this article that an economically driven vision of culture-led regeneration may serve to underestimate the diverse meanings which all social groups potentially invest in a development like that on NewcastleGateshead Quayside. It is in this context that Bianchini and Schwengel (1991) call for a genuinely public debate about the reimagining of cities, a debate that is not left to the marketing strategy and urban boosters, or which constructs an idealised middle class of what a city should be, but one that genuinely engages with the people that make a city what it is.

NewcastleGateshead Quayside

In order for a genuine public debate to take place and in order to understand the impact of culture-led regeneration and in particular the relationship between iconic projects and a sense of place and space, it is essential to contextualise the historical and sociological conditions under which such circumstances arise. This approach is not one that sits very happily with the short-termism associated with an approach to cultural policy that seeks to 'prove' the cultural case (Bailey et al., 2004). An approach that prioritises the meaning attached to iconic developments may prove far more beneficial in determining why a development is successful. This article will therefore focus on the findings that are beginning to emerge from a 10-year longitudinal project, the Cultural Investments and Strategic Impact Research (CISIR)[1] project funded by Gateshead Borough Council, Newcastle City Council, The Arts Council, England, One NorthEast and Culture North-East, which is concerned with the social, economic and cultural impact of cultural investment on NewcastleGateshead Quayside. Although just 3 years through its 10-year course, this research is beginning to indicate that iconic projects can serve a significant ideological function, at least if at the right place in the right time, as far as they play a key role in not simply reflecting a sense of local identity but in actually rearticulating and reconfiguring that identity in complex and paradoxical ways.

NewcastleGateshead Quayside has in recent years undergone a remarkable transformation. Millions of pounds of public and private investment have revitalised the Quayside both in the eyes of its people and, perhaps even more so, in the eyes of the outside world (Minton, 2003). This revitalisation centres around three iconic pieces of architecture: the BALTIC Contemporary Art Gallery built for £46 million; the Sage Gateshead Music Centre designed by Foster and Partners at a cost of £70 million and the Gateshead Millennium Bridge built at a cost of £22 million which in combination have served to redefine an area of industrial decline. The BALTIC is a new contemporary arts centre that overlooks the River Tyne. The Arts Council National Lottery funded project saw the conversion, by Gateshead Borough Council, of a 1940s grain warehouse into the largest gallery for contemporary art in the UK which aimed to attract 400 000 visitors annually. Originally conceived as an art factory, a place for artists from all over the world to work, the BALTIC has no permanent collection and boasts five generous spaces for contemporary exhibitions. Opened to the public in December 2004, The Sage Gateshead is not envisaged purely as a music venue. It is also a home for the Northern Sinfonia and Folkworks as well as a Music Education Centre. The reinvention of Gateshead Quay, which also includes residential developments and two international hotels, is linked to the Newcastle side of the Tyne by the Millennium Bridge, the world's first tilting bridge which was opened in September 2001 and won the RIBA Stirling Prize for architecture in 2002. In combination, these developments have given new life to NewcastleGateshead Quayside, providing the region with a renewed public focal point. It is, however, important to remember that the development of the Quayside has not been without its political tensions.

The history of the relationship between Newcastle City Council and Gateshead Council has not always been an easy one. The notion of NewcastleGateshead is in itself a construction of the destination-marketing agency Newcastle Gateshead Initiative, intent on cashing in on both the reputation of Newcastle upon Tyne as a regional capital and party city, and the cultural iconicity on the Gateshead side of the Tyne. The developments on the Quayside have undoubtedly played a key role in highlighting the potential benefits to be had from the two councils putting their local rivalry to one side for the common good. The Quayside has long provided a focal point for the region and, indeed, appears to be becoming increasingly important in this respect. However, the marriage between Newcastle and Gateshead is largely symbolic in nature and one issue this research will seek to address is the degree to which the renaming process is 'owned' by the people of Newcastle and Gateshead.

According to DCMS figures, the total of around £250 million investment by Gateshead Council on the Quayside in order to construct this world-class arts, leisure and residential development has in turn generated over £1 billion in private-sector funding. Given the public reception of the Quayside developments, common-sense would suggest that the NewcastleGateshead Quayside represents something of a success. In policy circles, NewcastleGateshead is often heralded as an example of the immense potential of investment of this kind (Minton, 2003). However, it would of course be grossly misleading to assume that the iconic nature of these developments guarantees success or that investment at a similar level will automatically kick-start regeneration elsewhere. There is, indeed, a body of work that questions the 'just add culture and stir' school of thought (Evans, 2001; Gibson and Stevenson, 2004; Jones and Wilks-Heeg, 2004).

Most importantly, the culture of a place is an essential ingredient to the success of culture-led regeneration (Jayne, 2004). The DCMS (2004, p. 22) itself recognises that the initial economic surge produced by a large project "can be difficult to sustain unless it is part of a wider regeneration and unless it is formally rooted in the community". The NewcastleGateshead example serves to illustrate how complicated such a relationship can be, not least in its construction of public space which frames

> a vision of social life in the city, a vision both for those who live there, and interact in urban public spaces every day, and for the tourists, commuters, and wealthy folks who are free to flee the city's needy embrace (Zukin, 1995, p. 259).

The relationship between iconic developments on the Quayside and the wider community lies at the heart of the CISIR project. The cultural dimension of the research programme has included a series of major surveys carried out by Market Research UK which seeks information on cultural values and attendance among the local population and how these factors relate to broader social and economic indicators on a national basis. However, the degree to which statistical data can inform our understanding of the actual meaning of culture-led regeneration is doubtful, not least because changes in attendance are often used to justify public funding in the arts.

In terms of measuring the apparent willingness of the population of NewcastleGateshead to take ownership of development on the Quayside, it is worth noting that CISIR respondents were advised that expenditure on the bridge, the BALTIC and the SAGE Gateshead amounted to £250 million about half of which came from public expenditure. Sixty-six per cent of NewcastleGateshead respondents in 2003 thought this was a reasonable amount, an insignificant drop from 69 per cent who felt the same in 2002. This compares with 27 per cent who felt this expenditure was too high in 2003 and 23 per cent in 2002—further evidence that the development has strong public support. Indeed, 95 per cent of respondents in NewcastleGateshead in 2003 felt that the Quayside was improving the national image of the area while 89 per cent felt that the developments were creating local pride in the area. But of course the above data do not in themselves prove anything. Granted, the CISIR project is beginning to unearth evidence that the Quayside is starting to have a significant impact on people's attitudes to culture (Bailey et al. 2004). However, as Evans (2004) suggests, it remains notoriously difficult to define and to quantify the social impacts of cultural activity. In order to delve beneath the surface of cultural investment, it is indeed necessary to address the meanings with which local people endow the Quayside and have endowed the Quayside over time.

A Centre of Urban Sociability

The Quayside is a space that is well accustomed to change. However, at its heart appears a long tradition of sociability. Over the remainder of the research programme, the meaning of the Quayside will be addressed from a variety of angles, but the first step along this path was constituted by a series of group interviews undertaken with older residents of Newcastle and Gateshead who were identified through Age Concern. The interviews were limited to older people in order to address the meanings attached to the historical development of the Quayside in as focused a fashion as possible. The aim of these interviews was to tap into the role of the Quayside as a key urban space and to address the meanings with which people have endowed that space over time. Questions of historical and cultural change are fundamental to the meanings that underpin people's relationships with the Quayside as the following quotations from older residents of NewcastleGateshead indicate. The Quayside has always been a social space

Everybody spoke to everybody down on the Quayside in the '40s. Perfect strangers. It didn't matter. But nobody did anything about the quayside. It was a disgrace! But, there was an excitement about it. It was a change to get out of the house.

You saw all life down there in all its stages!

For decades, the Quayside was a focal point for family outings. One Gateshead resident who used to live in Washington recalled how an annual visit to the Quayside was one of the most exciting events of the year

It was a great adventure. My family would be waiting to hear about what had gone on.

The Quayside was an industrious place but also a place characterised by comradeship and repartee: "[We] always enjoyed pay day because [we] got paid in the pub". The Quayside was a vibrant place, perhaps personified above all by Paddy's market which was the main attraction on a Sunday afternoon. Memories of the Quayside were overwhelmingly positive as were opinions about more recent developments: "And what is the BALTIC, the BALTIC then was a flour mill, but to me it is a tourist attraction now. It's lovely". It is indeed worth remembering that for a period the Quayside was, according to these respondents, pretty much derelict and unloved, a row of unremarkable warehouses and sheds—"In the 1960s I used to be a city guide and I used to be ashamed if it was my turn to go down by the Cooperage because it was such a filthy horrible site to show people". Things have changed considerably, many of the older people having visitors who expressly want to see the Quayside

> It is just a tourist attraction now because people want to go there. I send cards to my family in Canada with the bridge on and early on in the year I had my sister-in-law and my niece and they wanted to go down and see the bridge and wanted to go and see the BALTIC. My sister-in-law was not Canadian she was English as well and she had an idea of what it was like before she left and she wanted to go and compare and see what she thought of it.

> We went on a day trip. It made my day when I went down there and saw the bridge it was the first time I had been down and it just made my day the way the whole area had changed.

Even given their undoubted fondness for the past, for these older people the Quayside is symbolic of a new improved Newcastle-Gateshead

> [In those days] everything had to be loaded in the right way otherwise it would probably go down there or go down there you know but it had to be done in a proper manner. But as I say it is progress in a lot of ways but apart from feeling sad about not having this kind of thing now I do think it has improved and we are making use of the Quayside now and making money for Tyneside and Gateshead which is amalgamated now which they weren't at one time. It was dirty old Gateshead and rotten Newcastle it was.

The Quayside may have been 'dirty' and 'rotten', but that was in a sense irrelevant because it *belonged* to the people of Tyneside

> It was very dirty, it was just a dirty old hole, excuse me. But it was our Tyne you know. It was where Tyneside people were brought up. And they knew this.

In many ways, as it is perhaps today, the Tyne was a 'focal point' for the people of Tyneside. It was indeed, "the heart of Tyneside, the city grew up from there". It is under these circumstances that the Quayside retains its aura as a symbol of the north-east.

Culture-led regeneration does not inevitably lead to the construction of a 'blandscape'. Critics of developments on the Quayside in the 1990s may have been justified in describing the mixture of office, bars and restaurants on the Newcastle side in this fashion. However, in combination with the iconic projects across the river, the Quayside offers something very different. The Millennium Bridge, the BALTIC and the Sage Gateshead are symbols of the future rooted in the past. This is no better expressed than in the case of the BALTIC which was built in the shell of a disused flour mill. As Moore and Abbas (2004) and Forrest and Kearns (1999) suggest, the physical environment has an important role to play in fostering community morale and indeed for building bridges between generations and groups in a local community. The iconic projects on the Quayside provide an avenue through which this potential can conceivably be realised.

Behind the Quayside

It is not enough to say that investment on iconic projects on the Quayside feeds into the identity of the people in the region. What is it, under these circumstances, about the identity of the north-east that makes the Quayside developments work in the way they do? As Byrne (1999) indicates, the North's cultural identity is very much the product of the mixing of immigrant populations from Ireland, Scotland, Cumberland, Yorkshire, Scandinavia and other places in England who were attracted to the area by the prospect of high wages. The banks of the River Tyne once housed shipbuilding, chemical works, coalmining and other heavy industry that lay at the heart of the industrial might of the north-east (MacPherson, 1993). However, deindustrialisation brought with it urban decay as Power and Mumford (1999) point out. In discussing the Newcastle example, Power describes a situation in which Newcastle had entered a cycle of escalating physical decay in which houses were progressively being abandoned and boarded up. The causes of such a development, as Hall (2002) points out, were complex but characterised by the long-term structural decline of the economy, notably during the 1970s and 1980s, and thus long-term unemployment, with poor-performing schools perpetuating the problem, plus social disorder and even gang warfare. This was a particular problem in west Newcastle, despite signs that the city centre itself was by the turn of the millennium beginning to show signs of something of an urban renaissance with a thriving city centre

> attracting not merely tourists and night-time visitors but also now residents who were colonizing converted warehouses and new apartment blocks: urban renaissance and urban collapse were standing side by side, sometimes as little as a mile apart (Hall, 2002, p. 418).

Global circumstances and the deindustrialisation of the north-east created a set of circumstances in which regional particularity had to be transferred from production to consumption and this was an essentially divisive process (Vall, 1999). In this context, Keith Wrightson's thesis that Northern identity is about pride and truculence is insightful

> A northern upbringing frequently involves the inculcation of an unusually powerful set of attachments to place; a deep rooting in a particular physical, social and cultural environment. At the same time, however, those loyalties are strongly inflected, almost from the outset, by awareness of a questionable place within the larger social

and political geography of England (Wrightson, 1995, p. 29).

According to this view, pride qualified by anxiety breeds truculence. Perhaps it is in this way that cultural initiatives such as those we have described have apparently had such a fundamental impact on local peoples: the NewcastleGateshead Quayside gave the people of the region something tangible with which they could reassert their collective identities. The sociability generally associated with the people of Newcastle also plays an important role here. Newcastle, for instance, was recently voted one of the world's top 10 party cities by the Weisman Travel Agency. Lancaster points out that the working classes have been the 'leading' class in Newcastle for two centuries, the local élite having abandoned the city for the mansions of the Tyne valley. The end product of all this is a noisy and confident city and a city that is having to adapt to social, economic and cultural change: a city that fulfils many of the key requirements of successful city-making (Hall, 1998). As Lancaster puts it

> Cities never stay still, they are always changing, consciously or unconsciously trying to be something else. Cities are places where people strive to overcome the negative effects of past and current circumstances and struggle to create meaning, joy and hope in the place that history has located them (Lancaster, 1995, p. 7).

It is this sense that the emphasis needs to be placed on the relationship between individuals and their physical and social relationships, because it is this relationship that underpins the transactional nature of place. From this perspective, places are in a constant state of flux as the town 'rubs off' on its residents in a processual fashion. From this point of view, individuals actively construct and construe the experience of their immediate environment which is more than simply the product of broader cultural processes, but is about the relationship between people and place (Bonnes *et al.*, 2003; Twigger-Ross and Uzzell, 1996).

Many commentators struggle to grapple with the identities of spaces and places and how those identities are played out through history and NewcastleGateshead is no exception in providing a significant challenge to sociologists, geographers and historians alike (Minton, 2003). But the example of NewcastleGateshead Quayside also raises the possibility that investment in culture is not simply about regenerating the local economy, but can actually serve to revitalise the identities of the people of a city and even of a region; that it can provide new ways for those people to look into themselves and out of themselves. In other words, it can reinvigorate the relationship between cultural, place and personal identity and offer a permanent legacy. Such a realisation has significant implications for the ways in which policymakers engage with and indeed place expectations upon iconic cultural projects. As Hunt puts it

> The most successful cultural enterprises rightly announce themselves with an architectural statement, but they also draw on indigenous traditions which appeal to the city's self-identity. Yet all of them suffer from a common dependency upon lottery and state funds which ensures that so much cultural regeneration is dangerously dependent upon political fashion and consumer trends. Grand-standing, high-prestige developments funded by outside quangos usually falter if there is no local talent or support networks behind them (Hunt, 2004, p. 348).

In his book *The Uses of Disorder*, Richard Sennett (1970, p. 51) argues that in reconceptualising the city we should not be seeking to restore utopian visions of a small intimate urban sociability, but should rather seek to find "some condition of urban life appropriate for an affluent, technological era". Sennett argues that in an ever-elaborate bureaucratic and technological world, the social dimensions of urban life have rather been neglected

> There were hidden threads of social structure in ... poor city areas, threads that

give the people who lived there other regions of identity beyond their own poverty. Essentially, the last few decades of prosperity have righted the injustice these city people suffered, but at the cost of the breakup of their group life (Sennett, 1970, p. 53).

From this point of view, city life is less unpredictable and more coherent than it was in the past and, while this might be a good thing in terms of the efficiency of the city, it is not so good for us as human beings. Above all, a new centring on home and family has created a situation in which social spaces are conceived as intimate and small and therefore based around the home. In short, Sennett argues that the essence and diversity of urban life have been undermined leaving a situation in which our cities are crying out for new forms of complexity.

The argument being presented here is that yes, in some respects the iconic cultural developments represented on the Quayside and taken at face value are inevitably socially exclusive. The apartment buildings that have been developed immediately behind the BALTIC are more accessible to some social groups than others. At least some of the art presented in BALTIC is inevitably more accessible to some social groups than others. In many respects then, this project is inevitably one formed around the building-blocks of economic and cultural capital. However, those building-blocks can potentially produce new forms of complexity and diverse experience that may transcend this superficial exclusivity. Moreover, perhaps iconic developments such as that on NewcastleGateshead Quayside can successfully tap into and reconfigure aspects of place identity. Perhaps the Quayside will work because it offers a diverse range of new experiences, juxtaposing aspects of the arts, night-life culture and pride in place, that mean different things for different social groups and different identities. Lefebvre (1991) argues that the success of a city image depends upon the degree to which the physical image of a city and its rhetorical image complement each other.

Perhaps NewcastleGateshead Quayside will succeed in traversing the rhetorical to provide a new form of urban sociability.

Landmark sites such as that on Newcastle-Gateshead Quayside have a significant symbolic and material power. They make a powerful statement about a place and that place's intentions. But that statement is not, as we might assume, imposed upon the people of a city. Its meanings are at least potentially open to negotiation and it is the nature of that negotiation that researchers need to decipher if research into iconic culture-led regeneration is to teach us any genuine lessons. There is no one public space, as Zukin (1995) suggests. Urban space is experienced space and just because one space provides cultural opportunities that may appear to fit more readily into the habitus of a particular social group, does not necessarily mean to say it represents a form of oppression to another. It may indeed provide a means, however symbolic, of escaping from that oppression. Ultimately, landmark sights and in particular, waterfront regeneration schemes are the product of a complex of local, cultural, economic and historical factors (Bassett *et al.*, 2002). As Breen and Rigby argue

> Waterfront redevelopment and expansion is, in short, the best current example globally of the resilience of cities, of their ability to adapt to changed circumstances, to adjust to new technological impacts, to seize opportunities and to forge new images for themselves, as well as to create new or altered neighbourhoods for their inhabitants. ... Urban waterfront projects do not always succeed. But where they do, they have a dramatic and visible impact that is capable not only of enriching a city's economy but of improving its collective self-image (Breen and Rigby, 1996, p. 11).

Healey (2002) has also suggested that civic attention and thus cultural identity are drifting away from grand public plazas and architectural monuments. Nowadays, the football club or the retail precinct is the centre of civic attention. But the NewcastleGateshead example at least hints at the fact that this need not always be the case. One interpretation of the Quayside is as a centre of consumption, playing to the aesthetic sensibilities of the middle classes (Pollard, 2004). But that is one interpretation amongst many. It could equally be argued that global forms of consumption that appear on the surface to be imposed are actually renegotiated at the local level (Evans, 2001). It is in this sense that the Quayside has emerged as a focal point for the 'imagining' of Newcastle-Gateshead; an imagining that has developed into a mobilising force in the public realm of governance in Newcastle and Gateshead (Healey, 2002). Politically, the Quayside has been a catalyst for revitalising a climate of political collaboration between two rival councils. The challenge now is to maintain momentum; to use these iconic projects as a foundation upon which culture-led regeneration can undermine those aspects of social polarisation that are so often the inevitable consequence of post-industrial developments of this kind. But this is only the beginning of the story. If social polaristation is to be avoided, the iconography of the Quayside needs to precipitate a permanent legacy which taps into the cultural lives of all social groups. The BALTIC and the Sage Gateshead are at least vocal in their determination to appeal to a broad range of social groups, notably through Sage Gateshead's efforts to incorporate all forms of musical performance and through both organisations' education programmes. Whether success is achieved in this regard, only time will tell.

Conclusions

The meaning of 'culture' and the impact of cultural provision on place is in a sense intangible. Of course human beings endow places with meaning and thus identity is a sociospatial phenomenon (Neill, 2004). Liggett (1995, p. 252) therefore suggests that representational space is heavily loaded and deeply symbolic: calling upon shared experiences and interpretations at a profound level. From this point of view, iconic projects provide a key source of cultural meaning. Alternatively, Zukin (1991, p. 268) argues that urban space structures people's "perceptions, interactions, and sense of well-being or despair, belonging or alienation". In this context, I want to suggest that the Quayside, the oldest part of Newcastle and until the 19th century the commercial hub of the city, represents an especially important representational space for the north-east and thus plays a key role in structuring the above emotions. In many respects the Quayside has always been at the centre of the region, in terms of the region's industrial heritage, not least given the iconography of the Tyne Bridge opened in 1929. Thus, the contention that meanings can become more important than the facts in policy deliberation is a prescient one (Neill, 2004). But perhaps the key point here is that policy-makers and local people alike align themselves to imagined communities and in this case to an imagined post-industrial future. The Quayside offers the possibility of an optimistic future in an otherwise pessimistic age. However, the optimism engendered in such iconic developments is rooted in the foundations provided by NewcastleGateshead's industrial past.

The cultural identity of a place is not simply the product of the moment, but of the evolution and adaptability of time. For this reason, questions of identity should lie at the heart of the discussion of NewcastleGateshead, and also in discussions of culture-led regeneration more generally. This is a point taken up by Neill (2004), who refers to Hall, who in turn argues that

> identification is constructed through common origin and shared characteristics with people and groups, or perhaps with an ideal, and the solidarity that emanates from that ideal (Hall, 1996, p. 13).

In other words, identity is processual, marked by power relationships and uses a variety of cultural building materials from history, geography, religion, sexuality and so on (Castells, 1997). The construction of identity is as likely to be based on the symbolic as it is on the real: by imagined differences as compared with the other and by common characteristics shared by a particular social or indeed geographical group. Hall goes on to describe cultural identity as a sort of shared culture, a collective true self or common ancestry which may take precedence over other aspects of identity. For this reason, iconic developments cannot be understood in isolation. As Hayden et al. put it

> Restoring significant shared meanings for many neglected urban spaces involves claiming the entire cultural landscape as an important part of history, not just its architectural monuments (Hayden et al., 1996, p. 109).

The Quayside development is therefore a key ingredient in what Moore and Abbas (2004) describe as the yet unexplored symbiotic relationship between culture and place, but more specifically perhaps, the relationship between cultural *history* and space.

There is, of course, no straightforward answer to the question, can culture make cities work? The impact of cultural investment in iconic projects is highly site-specific. There is no magic formula for success. But the important point here and one that deserves further investigation is that despite the political nature of culture-led regeneration it does not necessarily produce a meaningless blandscape. Such a view represents an aesthetic simplification and not one that seeks to engage with the meanings with which people endow iconic projects such as that on NewcastleGateshead Quayside. I am not suggesting here that an approach to the impact of iconic projects in the urban landscape should be uncritical; far from it. But any such analysis should be steeped in the historical identities of people and places which can therefore provide a starting-point from which critical analyses can develop.

Developments on NewcastleGateshead Quayside emerged from a spirit in which politicians, policy-makers and Arts activists were determined to provide the region with the world-class facilities they thought it deserved. And yet Broudehoux paints a picture in which city leaders manipulate cultural forms and symbols to engineer consensus among city residents, foster local pride, and promote a shared sense of identity ... urban beautification also has a depoliticizing effect, and detracts attention from social and economic inequities by reducing the city to a surface assumed to be transparent and unproblematic (Broudehoux, 2004, p. 27).

In constructing such an image of iconic cultural development, Broudehoux presents a rather static image of city life and the meanings people attach to it. The iconic projects on NewcastleGateshead Quayside are landmark buildings that undoubtedly contribute to the pride and confidence of people in the region; an essential element to any programme of urban regeneration (Forrest and Kearns, 1999). But to describe this process as depoliticising underestimates the degree to which the meanings which people invest in developments of this kind are individualised and place-specific. As Zukin (1995) puts it, public space constitutes a window into a city's soul. NewcastleGateshead Quayside tells you as much about the north-east's industrial past as its ambitions for a post-industrial future. Although in its early days, the CISIR programme aspires to understand the degree to which that future can be a reality for the people of NewcastleGateshead and the north-east. The programme will continue to do so by seeking to analyse the way meaning is constructed around the Quayside, whilst comprehending the Quayside's economic impact in this context.

It is essential to seek out the motivations and expectations people bring to their interaction with cities in order to understand the likelihood that significant investment in iconic projects will succeed in specific places. Culture-led regeneration on the scale of NewcastleGateshead Quayside may indeed not work elsewhere, but at this time and in this place it offers a symbolic representation of a region that can succeed and a region that can begin to fight back from a period of industrial decline and neglect.

Vegara's thoughts on Bilbao are especially pertinent here

> The greatest miracle that Bilbao is experiencing is a dramatic change in attitude. The feelings of failure and pessimism brought about by prolonged economic crisis and political conflicts have given way to a collective optimism ... The majority of the Basque community—the public institutions, the private sector, and the civil society—is now convinced that it is indeed possible to reinvent Bilbao and the Basque Country in the new post-industrial age. This is the true miracle of Bilbao (Vegara, 2001, p. 94).

The degree of social and economic control that the south continues to exert over the north may or may not be exaggerated, but the fact that London's Millennium Bridge wobbled and Gateshead's did not is undeniably real. It is real for NewcastleGateshead as a city seeking to establish a sense of itself for consumption by the outside world, but most importantly it is real for the people who have lived all their lives in one or other of these two cities sitting either side of the Tyne.

Note

1. The CISIR programme of research is being conducted by colleagues at the Centre for Cultural Policy and Management, Northumbria University.

References

Augé, M. (1995) *Non-Places: Introduction to the Anthropolgy of Non-Places.* London: Verso.

Bailey, C., Miles, S. and Stark, P. (2004) Culture-led urban regeneration and the revitalisation of identities in Newcastle, Gateshead and the North East of England, *International Journal of Cultural Policy*, 10(1), pp. 47–66.

Balsas, C. (2004) City centre regeneration in the context of the 2001 European Capital of Culture in Porto, Portugal, *Local Economy*, 19(4), pp. 396–410.

Bassett, K., Griffiths, R. and Smith, I. (2002) Testing governance: partnerships, planning and conflict in waterfront regeneration, *Urban Studies*, 39(10), pp. 1757–1775.

BIANCHINI, F. and SCHWENGEL, H. (1991) Re-imagining the city, in: J. CORNER and S. HARVEY (Eds) *Enterprise and Heritage*, pp. 212–263. London: Routledge.

BONNES, M., LEE, T. and BONAIUTO, M. (Eds) (2003) *Psychological Theories for Environmental Issues*. Aldershot: Ashgate.

BREEN, A. and RIGBY, D. (1996) *The New Waterfront: A Worldwide Urban Success Story*. London: Thames and Hudson.

BROUDEHOUX, A.-M. (2004) *The Making and Selling of Post-Mao Beijing*. London: Routledge.

BYRNE, D. (1999) Is the North of England English?, *Northern Review*, 8, pp. 18–26.

CASTELLS, M. (1997) *The Power of Identity, The Information Age. Economy, Society and Culture, Vol. 2*. Oxford: Blackwell.

COWELL, R. and THOMAS, H. (2002) Managing nature and narratives of dispossession: reclaiming territory in Cardiff Bay, *Urban Studies*, 39(7), pp. 1241–1260.

DCMS (DEPARTMENT OF CULTURE, MEDIA AND SPORT) (2004) *Culture at the Heart of Regeneration*. London: DCMS.

DELANTY, G. (2003) *Community*. London: Routledge.

DICKS, B. (2003) *Culture on Display: The Production of Contemporary Visibility*. Buckingham: Open University Press.

EVANS, G. (2001) *Cultural Planning: An Urban Renaissance?* London: Routledge.

EVANS, G. (2004) Culture and regeneration: measuring impact, *Arts Professional*, 84, 4 October, pp. 5–6.

FLORIDA, R. (2002) *The Rise of the Creative Class*. New York: Basic Books.

FORREST, R. and KEARNS, A. (1999) *Joined Up Places? Social Cohesion and Neighbourhood Regeneration*. York: Joseph Rowntree Foundation.

GIBSON, L. and STEVENSON, D. (2004) Urban space and the uses of culture, *International Journal of Cultural Policy*, 10(1), pp. 1–4.

HALL, P. (1998) *Cities in Civilization: Culture, Innovation and Urban Order*. London: Weidenfeld and Nicholson.

HALL, P. (2002) *Cities of Tomorrow*. Oxford: Blackwell.

HALL, S. (1996) Who needs identity?, in: A. HALL and P. DU GAY (Eds) *Questions of Cultural Identity*, pp. 1–17. London: Sage.

HARVEY, D. (1990) *The Condition of Postmodernity*. Oxford: Blackwell.

HAYDEN, E., BORDER, I., KERR, J. *ET AL.* (1996) Strangely familiar: narratives of architecture in the city, in: M. MILES, T. HALL and I. BORDEN (Eds) *The City Cultures Reader*, pp. 109–110. London: Routledge.

HEALEY, P. (2002) On creating the 'city' as a collective resource, *Urban Studies*, 39(10), pp. 1777–1792.

HUNT, T. (2004) *Building Jerusalem: The Rise and Fall of the Victorian City*. London: Weidenfeld and Nicolson.

JAYNE, M. (2004) Culture that works? Creative industries development in a working-class city, *Capital & Class*, 84, pp. 199–210.

JONES, P. and WILKS-HEEG, S. (2004) Capitalising culture: Liverpool 2008, *Local Economy*, 19(4), pp. 341–360.

LANCASTER, B. (1995) City cultures and the 'Parliaments of Birds': a letter from Newcastle, *Northern Review*, 2, pp. 1–11.

LEFEBVRE, H. (1991) *The Production of Space*. Oxford: Basil Blackwell.

LIGGETT, H. (1995) City sights/sites of memories and dreams, in: H. LIGGETT and D. PERRY (Eds) *Spatial Practices*, pp. 243–273. Beverly Hills, CA: Sage.

LIVERPOOL CULTURE 2008 (2004) Fourth Grace gets £43m green light, (http://www.liverpoolculture.com/get-news-and-events/news_detail.asp?pid=20&ID=389; accessed 22 September 2004).

MACPHERSON, T. (1993) Regenerating industrial riversides in the north east of England, in: K. N. WHITE, E. G. BELLINGER, A. J. SAUL *ET AL.* (Eds) *Urban Waterside Regeneration: Problems and Prospects*, pp. 31–42. London: Ellis Harwood.

MARSHALL, R. (Ed.) (2001) *Waterfronts in Post-Industrial Cities*. London: Spon Press.

MCGUIGAN, J. (1996) *Culture and the Public Sphere*. London: Routledge.

MILES, S. and MILES, M. (2004) *Consuming Cities*. Basingstoke: Macmillan Palgrave.

MINTON, A. (2003) *Northern Soul: Culture, Creativity and Quality of Place in Newcastle and Gateshead*. London: DEMOS.

MOORE, J. and ABBAS, A. (2004) *Evaluating cultural regeneration: the role of place?* Presented at the *Conference of the International Association for People–Environment Studies* (IAPS 18), Vienna, July.

NEILL, J. (2004). *Urban Planning and Cultural Identity*. London: Routledge.

OWEN, J. (1993) The water's edge: the space between buildings and water, in: K. N. WHITE, E. G. BELLINGER, A. J. SAUL *ET AL.* (Eds) *Urban Waterside Regeneration: Problems and Prospects*, pp. 15–21. London: Ellis Harwood.

POLLARD, J. (2004) From industrial district to urban village? Manufacturing, money and consumption in Birmingham's Jewellery Quarter, *Urban Studies*, 41(1), pp. 173–194.

POWER, A. and MUMFORD, K. (1999) *The Slow Death of Great Cities? Urban Abandonment or*

Urban Renaissance? York: Joseph Rowntree Trust.

SENNETT, R. (1970) *The Uses of Disorder: Personal Identity and City Life*. New York: W. W. Norton.

SHORT, R. (1989) *The Humane City: Cities As If People Matter*. Oxford: Blackwell.

TWIGGER-ROSS, C. and UZZELL, D. (1996) Place and environmental processes, *Journal of Environmental Psychology*, 16, pp. 205–220.

VALL, N. (1999) Where are you from?, *Northern Review*, 8, pp. 27–34.

VEGARA, A. (2001) New millennium Bilbao, in: R. MARSHALL (Ed.) *Waterfronts in Post-Industrial Cities*, pp. 86–94. London: Spon.

WRIGHTSON, K. (1995) Northern identities: the longue durée, *Northern Review*, 2, pp. 25–34.

ZUKIN, S. (1991) *Landscapes of Power: From Detroit to Disney World*. London: California University Press.

ZUKIN, S. (1995) *The Culture of Cities*. Oxford: Blackwell.

Arts Festivals and the City

Bernadette Quinn

'What is a festival? It's something exceptional, something out of the ordinary ... something that must create a special atmosphere which stems not only from the quality of the art and the production, but from the countryside, the ambience of a city and the traditions ... of a region (de Rougement, quoted in Isar, 1976, p. 131; author's translation).[1]

1. Introduction

The past 15 years or so have seen a remarkable rise in the number of arts festivals in cities throughout Europe and elsewhere. Their growth has been such that it is now difficult to determine accurately the number of festivals in existence. Reasons explaining this proliferation lie in a series of interrelated factors that include changing approaches to urban management, structural changes in economic production, the use of culture as a means of restructuring wealth and job creation, and the unsettling effects of globalisation. All of these factors, in combination, have prompted a reconceptualisation of the festival as a useful strategy for the contemporary city to adopt in the attempt to reposition and differentiate itself in an increasingly competitive world. As Paddison (1993) explains, a city in pursuit of internal investment will compete with other cities through urban entrepreneurial displays. Festivals and events, as forms of entrepreneurial display, have come to be construed as vital elements in acquiring the investment needed for restructuring and regeneration (Robertson and Wardrop, 2004).

While the reasons explaining this recent proliferation are clear, the outcomes of cities' involvement with festivals are far less so. A number of researchers (Evans, 2001; Hannigan, 2003; Gibson and Stephenson, 2004; Richards and Wilson, 2004) have argued that, while cities use festivals and events with the intention of marketing themselves and creating place distinctiveness, the strategy may be counter-productive. Urban events, it is argued, run the risk of suffering from 'serial reproduction' (Richards and Wilson, 2004, p. 1932), of becoming formulaic (Evans, 2001) and hence devoid of any real connections with place. Yet, while the literature identifies in urban festivals the potentially homogenising effects of globalisation, other perspectives on culture-led urban regeneration argue that the reproduction of sameness need not be the outcome. Bailey *et al.* (2004), for example, argue that homogenisation is not inevitable, but is attributable to urban management approaches that fail to understand how local particuliarities could be cultivated to counter the globalising influences of cultural production in city arenas.

More generally, some have questioned the prevailing 'just add culture and stir' approach to urban regeneration (Gibson and Stevenson, 2004, p. 1), querying the extent to which it usefully serves public interests either in the short or the long term. A similar question could be asked of the role that festivals play in urban areas. Currently, the literature is very uncertain about their contribution. While there has been a lot of hype about the theoretically catalytic effect that festivals can have in terms of attracting visitors, spearheading the regeneration of derelict city districts and reclaiming public time and space for communal celebrations, hard evidence is in short supply. Evans (2001, p. 236) warns that contemporary festivals should be viewed as problematic if "their purpose and sustainability is of concern beyond the calendar cycle of ever-growing cultural feasts". This problem provides the focus for this article. It seeks to review the current state of knowledge about how festivals contribute to shaping the functioning of urban areas. It reviews existing literature on festivals and raises critical issues concerning the outcomes, particularly the non-economic outcomes, identifiable in the use of festivals in urban contexts. In suggesting ways of addressing emerging problems, the article suggests dialoguing with other literatures that have long sought to theorise the meanings historically associated with festivals and festivity.

2. Festival Meanings: Historical Perspectives

According to Turner (1982, p. 11), people in all cultures recognise the need to set aside certain times and spaces for communal creativity and celebration, and festivals have long constituted a vehicle for expressing the close relationship between identity and place. Ekman (1999), writing in a Swedish context, for example, described festivals as occasions for expressing collective belonging to a group or a place. In creating opportunities for drawing on shared histories, shared cultural practices and ideals, as well as creating settings for social interactions, festivals engender local continuity. They constitute arenas where local knowledge is produced and reproduced, where the history, cultural inheritance and social structures, which distinguish one place from another, are revised, rejected or recreated. To borrow Geertz's terminology, they can be said to represent an example of a 'cultural text' (Geertz, 1993), one of the many ensembles of texts that comprise a people's culture. Historically, interrogating festival settings has yielded insights into how a people's sense of their own identity is closely bound up with their attachment to place. In a European context, for example, Muir (1997) has written about the important function that public festivities played in towns across western Europe between the 12th and 18th centuries, those centuries during which civic consciousness, or the identification of individuals with their home town, came to be one of the distinguishing characteristics of European civilisation. Particularly in independent city-states, such as Venice,[2] public rituals and festivities were critical in consolidating civic identities in the face of internal division and external threats. Furthermore, such powerful city-states used festivities to exert control over their territories. Muir (1997) explains that Venice constructed and represented its colonial dominion through ritual, forcing subject cities to celebrate the feast days of St Mark, the patron saint of Venice. His analysis echoes Bonnemaison's comment on the 'hallmark event', which he says "functions like a monument, supporting and reinforcing the image of established power, whether religious or secular" (Bonnemaison, 1990, p. 25; quoted in Hall, 1992, p. 89). Both authors implicitly highlight the central role that power relations play in the reproduction of meanings in festival sites. This serves as a reminder that the construction of festival practices is intimately bound up with the cultural and social divisions that structure human population groups.

As important cultural practices, festivals have a long-established association with cities. It is thought that the first festival took place in Athens as long ago as 534 BC, in honour of the God Dionysos, the patron of wine, feast and dance (HOLND FSTVL, 2002). Then, as in all subsequent centuries up to the present era, festivals played important social roles both in public and private, religious and secular spheres. The forerunners of contemporary urban arts festivals can be traced back to the 19th century, to the Bayreuth Festival in 1876 and the Salzburger Festspiele in 1920. During the 19th century, as Bassett (1993) discusses in a UK context, the interest in cultural development was closely linked to both the growth of cities and the rise of urban élites. The festivals that emerged during this period tended to present programmes of high-quality classical works, interpreted by renowned performers within famous theatres or concert halls for the benefit of arts connoisseurs. There was no question but that the cultural forms and infrastructures promoted at this time were unambiguously concerned with the 'high arts'. Indeed, as Bassett (1993, p. 1774) argues, support for the arts was implicit in the efforts made by social élites to exert their dominance and demarcate social boundaries between themselves and the population at large. Festivals like those at Bayreuth and Salzburg contributed to the process of reaffirming the civilising and educational values of 'high' culture.

The post-war period witnessed an upsurge in the number of festivals being established. In an era where the drive towards reconstruction, political stability and the forging of international linkages through trade (including through a fledging tourism industry) set the tenor for economic and social advancement, the emergence of such nationally important festivals as at Avignon (France), Edinburgh (Scotland), Amsterdam (the Netherlands), Wexford (Ireland) and Spoleto (Italy) were important contributions to Europe's cultural infrastructure. While many of the leading arts festivals were based in major cities, several were not. In fact, sometimes, a more peripheral location, away from the culturally well endowed capital city functioned as a liberating stimulus for festival development. Frey (1994) has argued that, in countries with highly developed cultural policies (like Germany, Austria and Italy), festivals sometimes emerged as reactionary attempts to overcome the restrictions and inflexibility associated with established cultural institutions. This seems to have been the case at Avignon, where a desire to work away from the 'confines' of Paris was important for its festival's founder, the theatre director Jean Vilar (Isar, 1976). In other cases, festival initiatives have shown themselves to be highly reflective, as well as constitutive, of the resources, circumstances and people existing in particular places. They emerged in response to artistic needs lacking within that place and crystallised the key resources available there. Very often the human resource was of critical importance, with many festivals owing their existence to the commitment and vision of one or several key individuals.

Irrespective of location, these festivals introduced vibrancy at a time when much of continental Europe's cultural resources and architectural heritage lay in ruins. They often had international programming dimensions and international ambitions, with Edinburgh, for example, aspiring to be the 'Athens of the North' (Jamieson, 2004, p. 66) and Wexford seeking to position itself on the world stage (Quinn, 1998). The preoccupation was still with the 'high' arts and there was as yet little sign of any oppositional culture, although in southern France, the Avignon festival founded by Vilar in 1947 was rethinking the arts festival concept in an attempt to promote inclusiveness, accessibility and new forms of interaction between audience, artists and place. However, it was a pioneering initiative and its approach was as yet unorthodox. The challenge to dominant arts paradigms began to emerge more strongly in the 1960s and 1970s, when international student festivals at places like Zagreb and Nancy started experimenting with new artistic ideas and pushing out the boundaries of what was acceptable in terms of artistic production and performance (HOLND FSTVL, 2002). Festivals during these decades grappled with definitions of culture, challenging accepted definitions of 'high' and 'low' arts and gradually breaking down distinctions between the two. Festivals like those at Avignon and the Fringe at Edinburgh now operationalised this radical rethinking in their programming, their use of venues and in the ways in which they tried to engage audiences.

The forces exerting pressure for change among festivals, as in cultural arenas more generally, were part of a much broader movement seeking social change in tandem with the widespread economic restructuring being experienced throughout the Western world during the late 1960s and 1970s. These decades witnessed the emergence of grassroots social movements promoting a variety of causes such as anti-war, feminism, environmentalism and gay rights. They were driven by young people who shared a common interest in challenging prevailing norms and the existing social order. As Bianchini (1996, p. 4) has pointed out, these movements were often closely associated with 'alternative' forms of cultural production and distribution comprising, among others, free festivals, visual arts exhibitions in non-traditional venues and experimental theatre groups. To the fore in using festival production to achieve social aims was the Avignon festival in southern France. Under the innovative direction of Vilar, the concept of the festival here was being developed as something to be enacted with and through local and visiting populations, as opposed to something simply presented to them (Isar, 1976). The intention was that local residents, organisers, directors

and performers would effortlessly interact with each other and with their place, bringing it alive to the sounds and sights of music, dancing and art, in a spirit of festivity. To this end, festival events were housed not only in conventional venues but in the open-air, on streets and in squares as well as in cafes and restaurants. Events were programmed to happen at all times of the day and night. While the directorship of Vilar was not unique, it was certainly ground-breaking and inspirational for festival directors across Europe. It privileged the communal, participative dimension so central to the original concept of the 'festival', a word which derives from the classical Latin word *festum* meaning feast (Isar, 1976). As such, it signalled a move away from earlier attempts to use the arts festival, and the arts more generally, as a means of defining and maintaining social distinctions.

3. Festivals and Urban Policy: Evolving Recent Approaches

By the beginning of the 1980s, the national and international contexts shaping the role of cultural production in society were changing radically. Patterns of cultural consumption had evolved radically in recent decades with huge expansions in the consumption of mass media products in the home and a corresponding rise in diverse patterns of culture as well as leisure and tourism. Cities, as Zukin (1991) noted, were no longer functioning as landscapes of production but as landscapes of consumption. The collapse of the industrial base in numerous cities had prompted a serious search for alternatives and a shift towards the service economy. Many cities were beginning to see the logic in developing the kinds of cultural facilities needed to attract the skilled workers who would make up the new service class (Bassett, 1993, p. 1777). Simultaneous with this 'cultural turn' in the advanced industrial societies emerged a corresponding inflation of 'image production' (Zukin, 1998). 'Image production' or city marketing in the post-industrial era was as Ward (1998) stresses, an American invention, which in a European context tended to sit uneasily with a more intervensionist approach to urban governance (Ashworth and Voogt, 1994).

The dramatic expansion of festivals in urban areas since the late 1980s is explainable in respect of these changing circumstances. Their growth represents cities' attempts to use consumer-oriented, cultural forms to differentiate themselves in a highly competitive, increasingly global marketplace. As Waitt (2004, p. 403) reminds us, the concept of constructing festivals with the intention of drawing international attention to cities is not new. As long ago as 1859, the Handel Centenary Festival held in London's Crystal Palace, was marketed as a tourist attraction, with the organisers distributing some 50 000 prospectuses in the European offices of the railway companies serving the Crystal Palace (Adams, 1986, p. 18). The Holland Festival established in 1947 was designed to promote economic development in the Netherlands, while in 1953 the Irish Ministry for Industry and Commerce established 'An Tóstal', a cultural festival specifically constructed to attract tourists. Festivals have, however, taken on a new significance in the context of globalisation. They are now construed as entrepreneurial displays, as image creators capable of attracting significant flows of increasingly mobile capital, people and services. Major events are seen as being particularly effective in that they ally tourism objectives with urban planning (Roche, 1994), while simultaneously providing a means through which political and urban élites can refashion collective feelings of identity, emotion and consciousness (Cox and Wood, 1994). This civic boosterism line of thinking argues that major events generate in citizens a sense of pride and self-esteem (Mueller and Fenton, 1989). This sense derives in large part from the external affirmation that the event bestows on the city and the belief that through the event, the city is increasing its stature on the world stage. Accordingly, in spite of the increasing tendency for festivals to be constructed for consumption by privileged audiences, such as

visiting tourists or affluent locals, people in general tend not to oppose prevailing meanings amidst the 'showcase effect' of the event (Hiller, 1998). Meanwhile, the demand from growing numbers of increasingly mobile, experienced tourists continues to rise. No longer content simply to gaze (Urry 1990), the search for experiential holidays supports the widespread orientation towards a greater consumption of cultural goods and experiences, including festivals (Heinrich, 1988; quoted in Hall, 1992, p. 27).

4. Critical Perspectives on the Role of Festivals in Urban Policies

Fully aware of these evolving trends, several cities have invested heavily in festivals as part of their urban regeneration and city marketing strategies. As García (2004) points out, the level of investment in arts events does not rival that accounted for by major sports events (like the Olympics or the World Cup) or by major business showcase events (like Expos and World Fairs). Nevertheless, as part of the broader phenomenon that has seen an increasing use of the arts in urban regeneration processes, the rise of the urban festival has been very significant. The question specifically posed here is: what has this actually meant for the cities and city populations concerned? What roles has the arts festival played in advancing urban policy, contributing to urban life and facilitating the expression of cultural identities? Answering these questions is hampered by a profound lack of empirical research in the area. There has been a marked growth in the literature on festivals but this has tended to concentrate on economic outcomes and operational issues (Robinson et al., 2004; Richards and Wilson, 2004). In this context, it is difficult not to concur with Bailey et al. (2004) who suggest that the long-term social impact of culture-led urban regeneration remains something of a mystery. While a number of researchers have been interested in how the rise of the urban festival has impacted on the cities concerned, few if any comprehensive studies analysing festivals' contribution to urban policy have been undertaken. However, a number of researchers have written about the experiences of cities including Glasgow, Edinburgh, Galway, Barcelona and Sydney, and this literature has revealed a number of themes that further our understanding of the relationship between festivals and cities. These themes are discussed below.

4.1 The Festival as Image-Maker

Any attempt to analyse the role that festivals play in cities rapidly uncovers the very narrow manner in which cities tend to construe festivals. Above all else, city management's need to construct them as key elements of the city's place-marketing strategy predominates. Raising the city's international profile and attracting visitors seems to have become the *raison d'être* of the city festival. The emphasis is very much on the spectacular as opposed to any real consideration of process. It seems more a matter of style than of substance. As Evans put it

> Hard branding the city through cultural flagships and festivals has created a form of *karaoke* architecture where it is not important how well you can sing but that you do it with verve and gusto (Evans, 2003, p. 417).

Even when the festival or cultural event has been strongly culturally orientated at the outset, the city marketing impetus seems to overwhelm the process and crowd out any other potentially driving-force underpinning the use of festivals in urban arenas. The European Cultural Capital event (established in 1985) arguably illustrates this point. Its original aims were purely cultural, being concerned with achieving cultural expression, celebrating cultural diversity while simultaneously promoting unity among Europeans (Richards and Wilson, 2004, p. 1936). However, over time, individual cities have used the event to achieve different aims and often have used non-cultural measurements to calculate the event's effectiveness. Typically, it is the change in visitor numbers that is used to demonstrate success, as in the cases of Copenhagen (Fridberg and Koch-Nielsen, 1997) and Antwerp (Richards, 2000) or to predict success, as in Cork (Roche, 2004) and Liverpool (http://news.bbc.co.uk; accessed 6 April 2003).

While the concept of city marketing can be interpreted holistically to include social, cultural as well as economic aspirations (van den Berg *et al.* 1980), in practice the application of the concept in the US and in the UK has usually been limited to economic ideals (Paddison, 1993). Furthermore, by definition, city marketing strategies are duty bound to emphasise the attractive elements of place while simultaneously down-playing or diverting attention from less salubrious features. The festival, with its connotations of sociability, playfulness, joviality and community provides a ready-made set of positive images on which to base a reconstruction of a less than perfect city image. Add the 'arts' dimension to the festival, and another series of positive images are available for manipulating according to the positioning requirements of the city. It is not surprising, therefore, that many cities have seen in festivals a sort of 'quick fix' solution to their image problems.

Of the cities discussed in the arts-led urban regeneration literature, Glasgow has received particular attention. Designated as European Cultural Capital in 1990, the city is credited with using the event to overhaul its image as a depressed, problem-ridden, post-industrial city into an attractive and culturally interesting service-driven contemporary city. While this particular cultural event is held to have been critical in Glasgow's image revolution, the city already had a history of involvement with festival events since the previous decade. It supported the Mayfest festival in the early 1980s and progressed to involvement with a jazz festival in 1986 and the Garden Festival in 1988 (García, 2004). The bid for the European Cultural Capital therefore grew out of a context of commitment to cultural events and a belief in their ability to contribute to turning around a city's fortunes. Whether or not the 1990 Cultural Capital designation had sustaining effects on the city's image remains a subject for debate. While the city's image was boosted during the year itself, a study conducted by Myerscough (1991) suggested that it was not being maintained. A somewhat similar finding was reported in the case of the North of England's designation as 'Year of the Visual Arts' in 1996. The Arts Council of England (1997) reported that while the Year had succeeded in raising awareness of the arts in the region, a more sustained arts programme would be needed if longer-term attitudinal change was to be achieved. Nevertheless, in the Glasgow case, media commentary (Ryan, 2002) and local tourism agencies (GGCVTB, 2002) continue to applaud the positive image transformation.

4.2 The Festival as Tourist Attraction

Glasgow, with greater certainty, can point to the increase in visitor arrivals that the 1990 designation stimulated. The year before Glasgow made its bid in 1983, visitors to the city numbered 700 000. By 1990 they numbered 3 million, of whom 600 000 were attributable to the event (Paddison, 1993). By 2002, they numbered 4 million annually, by which time the city had become Europe's fastest-growing conference destination (Ryan, 2002). Undoubtedly, the ambition of generating substantial tourist flows is a priority for many of the cities that use festivals as part of their urban regeneration/city marketing strategies. The European City of Culture examples cited earlier are cases in point. There is now an extensive tourism literature treating the rise of festivals as tourist attractions. The use of the term 'festival tourism' is increasing among tourism researchers, the vast majority of whom conceive of the festival primarily in terms of its economic potential. Increasing numbers of empirical studies capture the growing reality of city managers devising/reconstructing festivals as economic catalysts and report gains in terms of numbers of visitors attracted, revenue generated, extension to the tourist season and columns of media inches generated. Far fewer studies hint at, much less problematise, the narrowness of vision inherent in such approaches. Some researchers acknowledge that several of the festivals currently functioning as key city marketing strategies initially evolved from a more rooted connection with either place, community, tradition and/or art form. Yet while this has many implications (Hughes, 2000), these are rarely treated in the literature. Robertson and Wardrop (2004, p. 123) for example, point to the importance of acknowledging that the Edinburgh festival was not established with tourism objectives in mind yet fail to elaborate on this point. They go on to suggest that some involved in festivals and events have historically held that applying strategic frameworks to festivals and events serves only to stifle entrepreneurship and creativity. However, this comment is left without further development.

A contrasting set of perspectives can, however, be found among researchers who locate their enquiries within different domains. A well-established line of social science enquiry into festive practices has paid considerable attention to commodification, authenticity and cultural identity issues. Here, the extent to which the presence of tourists negatively affects festival processes is a key question (Boissevain, 1992; Sampath, 1997; Hitchcock, 1999). There has been a school of thought that associates tourism with the inevitable debasement of cultural meanings (see, for example, Greenwood, 1976, 1989), with Greenwood (1976, p. 141) arguing that it is "only the local people who have learned about the 'costs' of tourism. The outside investors and the government have been reaping huge profits and are well satisfied". However, this view is countered by the assertion that change is inevitable and has always been integral to the reproduction of festivities. Cohen (1988), for example, argues that through commoditisation, cultural entities can assume new meanings for the producers. This more nuanced interpretation suggests that

> rituals that may have been meaningful in the past for an internal public can evolve, under the influence of tourism, to become a culturally significant self-representation

before an external public (Cohen, 1988, p. 382).

Theoretically, an important argument implicit here is that local residents, as producers and as established audiences, can engage meaningfully in festivals in ways that address both their own needs as well as those of visitors at the same time. Empirically, however, evidence to support this theoretical position is scarce, and from artistic and broader cultural perspectives, the merits of engaging with tourism remain highly debatable.

A symposium of festival producers, policymakers and directors in the Netherlands (HOLND FSTVL 2002), for example, spoke of festivals leading to a 'tourist invasion', as at Oerol (Netherlands) and Avignon (France), where places come to "look more like a holiday resort than an arts festival" (HOLND FSTVL, 2002, p. 11). Landry *et al.* (1996) argue that, when a festival focuses on external audiences, a result can be to limit the ability of the artists to question, challenge and criticise. At such points in the festival's evolution, critical questions arise as to its ability to retain artistic autonomy, to maintain the quality of experience in the face of increasing pressures to commercialise. Such questions are debated in Quinn's (2005) analysis of the Galway Arts Festival. Set in the western seaboard city of Galway (population of 67 000; Central Statistics Office, 2001), this combined arts festival was established to celebrate and promote appreciation of the arts in the region in 1978. Informed initially by socially inspired objectives, the festival has expanded greatly from modest beginnings to become one of the largest and most popular festivals in the Republic of Ireland. The festival's growing popularity as a tourist attraction for both national and international audiences and its growing commercialism have created a series of dilemmas for the festival. Since the mid 1990s, it has been grappling with the tensions posed by trying to balance deep-rooted, socially aligned artistic goals on the one hand with often conflicting economic imperatives on the other. Adapting to changing circumstances is not an easy process. The Galway Arts Festival remains committed to serving local needs but the city's rapidly changing social and economic contexts mean that both production and consumption contexts have changed. Developments in the city's cultural infrastructure, partially achieved through the efforts of the festival, have facilitated a more commercial approach to arts production, for example,. Meanwhile the city's rapidly growing population, in both residential and tourist terms, is creating an unprecedented breadth and diversity of artistic needs.

For local people, these changes are all too evident and survey data gathered among city residents demonstrated clear concerns that the festival was losing its meaning for local people in the face of pressures to internationalise both its programme and its audiences from such gatekeepers as the national media, the state tourism agency and major sponsors. There was a clear sense that the festival's growing commercialism is problematic for local people. Increasing professionalism and the growing stature of the festival both as commercial enterprise and as tourist attraction underpinned a perception that the festival is becoming increasingly exclusive and inaccessible.

Yet, it is the very possibility of involvement and participation, and the potential to challenge, re-order, subvert and disrupt, that social scientists and others have held to be inherent in the concept of festivity (Willems-Braun, 1994; Ozouf, 1988). Social theorists have construed festivals as liminal, 'time out of time' spaces (Bahktin, 1984), replete with possibilities for challenging social conventions, social order and authority, and inverting society's cultural norms. Cultural commentators conceive of festivals as risk-takers, as opportunities to challenge the *status quo* and push out boundaries. Within the prevailing conceptualisation of arts festivals as city marketing strategies, festivals are permitted little scope for unlocking such potential, potential that would seem to be unlimited. Of even greater concern is the possibility that, conceived of as such, festivals may be both compounding the social difficulties that necessitate renewal and regeneration programmes in the first place and heightening tensions in already contested arenas. In the light of existing research knowledge about urban arts festivals, it does not seem unreasonable to ask whether city authorities have even begun to exploit the potential of arts festivals. The use of culture only for marketing purposes is limiting (Landry *et al.* 1996) and the broad-ranging conceptualisations of festivity evident in the literature contrast sharply with the tangible but narrow construction of festivals merely as economic generators.

4.3 The Festival as Community

As Landry *et al.* (1996) argue, the crude interests of the local economy and of the city as a whole do not always coincide. There is often a failure to appreciate that image campaigns with little grounding in local needs and aspirations can backfire. This is because genuine festivals must be "rooted on society, in real life" (Isar, 1976, p. 126), while arts festivals, if they are to be "artistically responsible festivals" (Degreef, 1994, p. 18) must respond and evolve in tandem with the changing artistic needs felt by diverse resident and visitor community groups within a place. These concepts can become lost in the face of increasingly persuasive city branding. Hannigan (2003) recently introduced a series of articles that highlight the rising influence of commercial marketing strategies in the shaping of the contemporary city. His comments emphasised growing concerns about consumer democracy, the integrity of public space and the cultural diversity of the metropolis. In theory, festivals could have a role to play in countering the social crises faced by cities in the context of globalisation. Hughes (1999) suggests that the growing interest in festivity in the 1990s is linked to its use as a social strategy to combat the growing alienation and insecurity felt in public space. Indeed, festivals have been historically construed as mechanisms through which place-based communities express identities, celebrate

communally held values and strengthen communal bonds (Jackson, 1988; Marston, 1989; Smith, 1996). In reality, however, how do these functions fare under the prevailing entrepreneurial approaches to urban management? (Putnam, 2001; Waitt, 2004). According to Zukin (1998), strategies of urban redevelopment based on consumption focus on visual attractions that make people spend money. Festivals as visual attractions require willing participants in the guise of spectators as well as consumers. Yet, what of more participative, communal engagement? Nurse (1999) argues that the Caribbean carnivals, now found in almost every city in North America and Britain, represent good examples of festivity in action. Spread along diasporic networks to globally disparate communities with Caribbean kinship ties, these carnivals constitute the largest event in terms of attendance and economic activity generated in their respective host cities (Nurse, 1999). Yet, the substantial tourism and economic activity dimensions do not overshadow the profound social meanings of these festivities. For diverse groups living within and attracted to these carnival cities, the carnival maintains its status as a hybrid site (Bhaba, 1994) where cultural identities, notions of belonging and values systems are celebrated, contested and negotiated. Evans (1993; quoted in Evans, 2001) concurs, arguing that major cultural events such as the carnival *Mas* in Trinidad and their diasporic reproductions in Toronto (Caribana) and London (Notting Hill) still retain indigenous involvement and strong shades of their original purpose. There is nothing inevitable about this sort of rooted cultural reproduction. Rather, it reveals the nurturing of specific broad-based objectives and deliberate efforts to achieve certain outcomes. As Owusu and Ross (1988) discuss, such carnivals actively seek to combine local, participant and tourist, and take months of planning, workshops, craft production and rehearsal before the events themselves take place.

In general, however, there is often little sense of collective or social responsibility evident in the contemporary promulgation of festivals in urban areas. Garcia (2004, p. 108) argues that two widely criticised aspects of Glasgow's 1990 European Year of Culture event were its failure to promote the involvement of geographically peripheral and socially deprived communities in arts activity, and its inability to act as a platform for representing local cultures. According to McLay (1990), Glasgow's Year of Culture acted as a 'superficial make-over', focusing on the privileged few while 'covering up' the real concerns of the city's working-class majority. Paddison (1993) similarly asks whether the 'new Glasgow' emerging through this cultural reconstruction process had any relevance for those experiencing poverty on the city's margins. Somewhat relatedly, Jamieson's (2004) research offers critical perspectives on the Edinburgh Festival. 'Edinburgh, the Festival City', with its numerous annual and contiguous summer festivals, is widely accredited with stimulating major tourist flows (Scotinform, 1991), generating some £125 million expenditure, sustaining nearly 4000 jobs across Scotland (Edinburgh International Festival, 2001) and creating a strong city image (Prentice and Andersen, 2003). Jamieson moves beyond the economic indicators to reveal a festival city that is spatially structured in a way that privileges visiting festival audiences and contains them within parts of the city deemed 'appropriate' for cultural consumption. Meanwhile, according to Jamieson (2004), the city's outskirts, where Edinburgh's socially deprived reside, remain relatively free of festival activity. This structuring of the festival, she argues, serves both to reassure cultural tourists of a safe encounter with the city and to marginalise those Edinburgh residents living in its peripheral housing estates. In consequence, there is little possibility of the festival engaging in "processes that might disrupt the social construction of the festival or challenge the dominant social meanings being reproduced therein" (Jamieson, 2004, p. 72). It is tempting to argue that the type of festival landscape discussed by Jamieson is associated only with 'high art' festivals. Waterman (1998, p. 66) asserts that high-brow festivals still explicitly prefer to

present themselves as élitist, citing the case of the Israel Festival as one that is unashamedly élitist, directing itself at the few who can afford to attend and believing that it should not be denigrated for this. However, the reproduction of difference through festival practices can equally characterise festivals more commonly perceived to be populist and accessible. Quinn (1998), for example, found that in the case of the Galway Arts Festival, the inclusive image of the festival belied a more complex agenda underpinning the construction of the festival. In this case, high art forms like classical music and more traditional forms of theatre involving the Irish language, and their devotees, were marginalised in favour of the more popular music, film and street theatre events that have become the festival's hallmarks. Similarly, the reproduction of difference, irrespective of the power dynamics at play, will in all likelihood encounter strategies of opposition or resistence (however passive) from those groupings who feel themselves to be excluded from the key decision-making processes (Quinn, 2003). It is such processes that at least partially account for the emergence of fringe/unofficial/alternative festival offerings in places like Wexford, Edinburgh and Avignon.

More generally, however, the implication in the literature is that what is often consumed and experienced in festival settings is an idealised, sanitised version of the city where real opportunities for genuine engagement with the culture and multiple realities of the place, for both local and visiting populations remain sidelined. The examples cited above echo Judd's (1999) critique of Disneyfied 'Latin Quarters' and their associated festivals as 'islands of pure consumption' for visiting populations. Such 'islands' are more likely to contribute to racial, ethnic and class tensions than to an impulse towards local community (Judd, 1999, p. 53). The centuries-old Lenten carnival celebrations in Venice, revitalised in the late 1980s with a tourism remit, provide an insight into this problem. In tourism terms, the contemporary Venice carnival is a magnet attraction, attracting some 900 000 visitors to the city over a 12-day period (E. Busolin, Commune di Venezia; personal interview, 2002). During peak carnival periods, it is no exaggeration to say that the city becomes overwhelmed. Bridge closures and one-way pedestrian systems are often needed to maintain mobility and many citizens in this ageing city choose to stay at home because they cannot negotiate the crowds. If mere visitor numbers were indicators of success, the Venice carnival would be impressive. However, while it extends the city's tourist season into the early part of the year (the carnival typically takes place in early February) and thus achieves one of its core revitalisation objectives, it is arguable whether the carnival contributes anything to reducing the city's detrimental overreliance on the excursionist market. Meanwhile, from a local resident's perspective, it is clear that the carnival is not meeting local needs. Even the 2002 carnival director, Fabio Momo, accepted that "the people of Venice cannot stand the carnival anymore". He elaborated further by saying

> the local becomes a spectator and isn't a participant anymore. But the carnival is a party and you cannot be a spectator in a party, you have to participate (F. Momo, Director of Carnevale di Venezia; personal interview, 2001).

Data derived from surveys undertaken among a representative sample of the local population during the 2002 carnival revealed widespread dissatisfaction with the current reproduction of carnival and suggestions for change revolved around three key areas: more spontaneity in the programming, including more events in outdoor public spaces, increased participation for locals; and, tighter controls on the presence and movements of visitors in the city during carnival (Quinn, 2004). The 'Friends of the Venice Carnival Association' founded in 1987 and chaired by Guido Rossato is very aware of declining local participation. In an attempt to counter the problem, it organises courses, free of charge, where local people can learn how to make inexpensive carnival costumes

and be encouraged to take part in Carnival activities in greater numbers. At the same time, however, local people's declining participation in, and negative attitudes towards, the contemporary carnival in no way imply a disinterest in this deep-rooted cultural practice. On the contrary, survey data identified a deep attachment to, and recognition of, the historical tradition of civic ritual that has been such a signifier of Venice for centuries.

In considering ways in which to reignite collective endeavour and restore civic engagement, Putnam (2001) asks that we consider increasing participation in, rather than consumption and appreciation of, cultural activities. He also suggests that we "discover new ways to use the arts as a vehicle for convening diverse groups of fellow citizens" (Putnam, 2001, p. 411). This is easier said than done. Local respondents interviewed as part of the Venice study could easily point to solutions. However, these are unlikely to be implemented or even made known to the carnival organisers because of the conflicts of interest at issue. As mentioned before, festivals are not natural occurrences, they are social constructions that bear heavy signs of authorship. In the case of the Venice carnival, tourism-dominated business interests prevail.

4.4 Globalisation and Local Diversity

Despite the role of agency and the politics inherently involved in the negotiating and challenging of meanings reproduced in festival settings (Jackson, 1988; Marston, 1989), a recurring charge in the literature is that originality is often replaced by imitation. Far from adding to place distinctiveness, the proliferation of festivals is at least partially explained by a formulaic approach to duplicating festivals found to have been 'successful' in particular city contexts. This is part of the phenomenon identified by Scott (2000), who writes about a deepening tension between culture as something grounded in place and culture as a pattern of non-place globalised events and experiences. In response, arts commentators have been prompted to deplore the dilution of quality, originality and difference. Clark is not alone in arguing that

> The modern festival ... is a sort of supermarket where the paying public is persuaded to bulk-buy processed culture [and that] such events quickly start to look the same (Clark, 2004, p. 34).

Any assessment of the role played by globalisation processes in reproducing sameness should consider that festivals have long had an outward orientation. They have been always characterised by 'interrelations' rather than 'autonomy of place', to use Hannerz's (1988) terminology. From a purely artistic perspective, they serve as forum for exchanging and comparing experiences and ideas, and for prompting collaboration with other arts festivals and practitioners. From a broader social perspective, they serve as vehicles through which cultural meanings are expressed for interpretation both by the place-based communities themselves and by the outside world. Thus, current problems concerning the dilution of originality and difference are not inevitable. Rather, as the performing arts literature would argue, they result from a failure to acknowledge that, while festivals are in part produced by and through globalisation processes, they can simultaneously respond to the challenges posed therein (Klaic et al. nd, p. 30). From a purely artistic perspective, this requires a finely balanced approach. The danger facing internationally oriented festivals is that they may neglect their local resources and cultural needs in the process. Building relationships with artists, audiences, with the local artistic environment, as well as with business, external audiences and with the media, is critical to the effective functioning of an arts festival (HOLND FSTVL, 2002). Festivals

> have to augment the existing supply and the local infrastructure. They have to add something .... that would not exist without the festival (HOLND FSTVL, 2002, p. 18).

The Symposium organised by the Holland Festival in 2002 offers the LIFT festival in London as an example of a festival that reflects on its own role, on its contribution to local cultural supply and reconfigures itself accordingly (HOLND FSTVL, 2002). This degree of self-analysis, also evident in the Galway Arts Festival (Quinn, 2005) is necessary if festivals are to maintain integrity and continue to contribute in a meaningful way to the social and cultural contexts in which they operate. Another festival offered as an example of one that manages to balance the needs of different stakeholders is the Malta Festival in Poznan, Poland. This festival attracts people from the area in addition to domestic visitors from elsewhere in Poland and professionals from abroad. A key part of its success is its programming to attract different audience groups simultaneously (HOLND FSTVL, 2002).

More fundamentally, structuring festivals such that they connect with, but are not overwhelmed by, globalisation processes requires a much deeper consideration of the social and cultural particularities of the cities in question. Bailey et al. (2004) argue that the future could lie in viewing cultural planning as being about engaging with the lives of those people who live in the city rather than being about regenerating the city itself. In their view, cultural forms of consumption can actively enhance and enliven communities. They do so "because culture matters for its own sake and not merely as a means to an economic end" (Bailey et al., 2004, p. 64). According to Zukin (1998, p. 836) cities have begun to view the increasing multi-ethnicity of urban populations as a source of cultural vitality and economic renewal. The diversity afforded by multiple cultural practices and value systems is construed as a series of opportunities that can be cultivated to strengthen the city's overall appeal and distinctiveness. However, cities must move beyond a preoccupation with image-making. Landry et al. (1996) highlight the case of Bradford city as one which has successfully involved local ethnic communities. The Bradford Community Festival was started in 1987 by the Economic Development Unit of the City Council and runs over three weeks. Its objectives were that the whole community irrespective of origin, geography or interests, would be involved and would find a voice through the festival. The Mela was launched as part of the festival programme in 1988. Since then, the popularity of the Mela has increased such that by 1996 it had approximately 140 000 visitors. Landry *et al.* (1996, p. 42) concluded that, while the Mela has an undoubted impact on tourism in Bradford, its greatest value is the fact that it has succeeded in stimulating significant involvement among the local Asian community. In contrast, Garcia (2004), in an analysis of the Olympic Arts programme hosted in Sydney in 2000, reported a series of criticisms from local arts groups representing cultural minorities regarding the event's failure to capture the diversity of contemporary Australia. She concluded that the Olympic Arts Festivals were not able to provide authentic cultural experiences, but rather were exotic commodities for the enjoyment of visitors and White locals (Garcia, 2004, p. 110).

5. Conclusion

Now it's festivals, festivals everywhere. Big ones, small ones, wild ones, silly ones, dutiful ones, pretentious ones, phony ones. Many have lost purpose and direction, not to mention individual profile. Place a potted plant near the box office, double the ticket prices and—whoopee—we have a festival (Bernheimer, 2003).

As Bailey *et al.* (2004) argue, cultural forms of consumption can actively enhance and enliven local communities. However, there is a strong sense in the literature that festivals, as examples of cultural forms of consumption, are not managing to achieve this end. The crux of the problem appears to lie in the failure of cities to acknowledge the critical importance of understanding and responding to the needs of local places, and of closely linking city marketing and urban regeneration strategies with the specificities of particular city contexts. Arts festivals are conceived of in far too narrow a vein. City authorities seem

to misunderstand the social value of festivals and construe them simply as vehicles of economic generation or as 'quick fix' solutions to city image problems. However, the tasks of conceptualising the problems at issue and devising appropriate policies are hampered by the scarcity of empirical research conducted in the area. The literature is replete with passing references to the social and cultural value of arts festivals, but there is a real shortage of in-depth, empirically grounded analyses of the issues involved.

Could not urban policy-makers be persuaded to conceive of arts festivals in terms of quality of life, cultural and social outcomes and not simply in terms of their economic and image creation outcomes? Such a shift would of course require a more long-term, holistic approach to city management. It would also require a much more integrated, consultative approach. Garcia (2004) has already identified a lack of co-ordination between event organisers, tourism bodies, city planners and the arts community as a key difficulty. An obvious void in the literature is the absence of any sort of dialogue between those who theorise about arts festivals within performing arts and theatre studies domains and those who strategise around them in urban planning contexts. It is insightful to note how the same subject is conceived of in completely different ways. The way of thinking and even the language used in these different literatures are starkly contrasting. Central ideas in the arts literature on festivals include: that festivals be 'artistically responsible' (Degreef, 1994, p. 18); that they respond to specific artistic needs genuinely felt within their place; that they be conscious of the need to add to the regular supply of arts provision existing on a year-round basis; that they dialogue with their diverse constituents and reflect on their social and cultural functions. Klaic *et al.* infer a strong recognition that festivals are not simply artistic entities, but that they can be implicit in local development and urban regeneration processes. They suggest that a festival

enables the residents to create a new vision, a way of looking at the place where they live from another point of view. It can improve the quality of communication among the residents and enhance the mutual understanding of social, ethnic, age and cultural groups (Klaic *et al.*, nd, p. 48).

All of this helps to create/reinforce the self-confidence of residents and change the perception of the area within and outside the community. This, they argue is "an essential step in any process of urban regeneration" (Klaic *et al.*, nd, p. 48).

The strong emphasis on understanding the place and the communities who live within that place is striking, and is mirrored in reverse measure by the extent to which it is often absent within the city marketing literature. Bakhtin's (1984) theories on carnival, as well as depicting ambivalence, challenging binary oppositions and circumventing regular social structures, also posit the people involved as both actors and spectators. Festivity depends on multiple forms of engagement. Too often contemporary urban arts festivals envisage only spectating roles for local residents and this strongly dilutes the cultural meanings that could be promulgated. Too often also, they fail to acknowledge the multiple realities and conflicting meanings that can be hidden beneath their image-conscious 'stage-managed' veneers.

In conclusion, given the store of empirically grounded literature currently available on arts festivals in city contexts, it is difficult not to agree with Bailey *et al.*'s (2004, p. 47) assertion that, when the democratisation of culture and the empowerment of local communities are cited as outcomes of culture-led regeneration, the rhetoric is drawing on mere assumptions. Urban policy-makers need more than assumptions to inform decision-making. If arts festivals are to achieve their undoubted potential in animating communities, celebrating diversity and improving quality of life, then they must be conceived of in a more holistic way by urban managers. Researchers have a role to play in investigating the multiple ways in which festivals realise such potential. Ultimately, as Landry

et al. (1996) stress, cultural investment can only do so much. Equally, arts festivals can only achieve so much. However, it seems abundantly clear that the growing investment evident in urban arts festivals in recent times is not yielding optimal returns. Until prevailing conceptualisations of festivals acknowledge their latent social and cultural potential, this will remain the case.

Notes

1. 'C'est quoi une festival? ... C'est d'abord une fête. C'est quelque chose d'exceptionnel, qui sort de la routine ... et qui doit créer une atmosphère spéciale, à laquelle contribuent non seulement la qualité des oeuvres et de leur exécution, mais le paysage, l'ambiance d'une cité et la tradition ... d'une région.
2. Venice remained an independent city-republic until 1797.

References

ADAMS, R. (1986) *A Book of British Music Festivals*. London: Robert Royce Ltd.

ARTS COUNCIL OF ENGLAND (1997) *Visual Arts UK: public attitudes towards and awareness of the Year of Visual Arts in the North of England* (Summary). ACE Research Report No. 15, December.

ASHWORTH, G. and VOOGT, H. (1994) Marketing and place promotion, in: J. GOLD and S. WARD (Eds) *Place Promotion: The Use of Publicity and Marketing to Sell Towns and Regions*, pp. 39–52. Chichester: Wiley.

BAHKTIN, M. (1984) Rabelais and His World, trans. by H. Iswolsky. Bloomington, IN: Indiana University Press.

BAILEY, C., MILES, S. and STARK, P. (2004) Culture-led urban regeneration and the revitalization of identities in Newcastle, Gateshead and the North East of England, *International Journal of Cultural Policy*, 10(1), pp. 47–65.

BASSETT, K. (1993) Urban cultural strategies and urban regeneration: a case study and critique, *Environment and Planning A*, 25 pp. 1773–1788.

BBC *News* (2003) Liverpool named capital of culture (http://news.bbc.co.uk; accessed 6 April, 2003).

BERG, L. VAN DEN, KLAASSEN, L. H. and MEER, J. VAN DER (1980) *Marketing Metropolitan Regions*. Rotterdam: European Institute for Comparative Urban Research.

BERNHEIMER, M. (2003) Beyond the Big Three, *Financial Times*, 28 June, p. 2.

BHABA, H. (1994) *The Location of Culture.* London: Routledge.

BIANCHINI, F. (1996) Culture, economic development and the locality: concepts, issues and ideas from European debates, in: S. HARDY ET AL. (Eds) *The Role of Art and Sport in Local and Regional Economic Development*, Proceedings of the Regional Studies Association Annual Conference, November 1986, pp. 2–10. Regional Studies Association.

BOISSEVAIN, J. (Ed.) (1992) *Revitalizing European Rituals.* London: Routledge.

BONNEMAISON, S. (1990) City politics and cyclical events, in: *Celebrations: Urban Spaces Transformed, Design Quarterly*, 147, pp. 24–32. Cambridge, MA: Massachusetts Institute of Technology for the Walker Art Center

CENTRAL STATISTICS OFFICE (2001) *Census of Ireland.* Dublin: Central Statistics Office.

CLARK, A. (2004) High brow, low blows, *Financial Times Magazine*, 19 June, pp. 34–37.

COHEN, E. (1988) Authenticity and commoditization in tourism, *Annals of Tourism Research*, 15, pp. 371–386.

COX, K. and WOOD, A. (1994) Local government and local economic development in the United States, *Regional Studies*, 28, pp. 640–645.

DEGREEF, H. (1994) European festivals: confrontations of cultural establishment and popular feast, *Carnet*, 2, pp. 16–22.

EDINBURGH INTERNATIONAL FESTIVAL (2001) *Annual Report.* Edinburgh International Festival, Edinburgh.

EKMAN, A. K. (1999) The revival of cultural celebrations in regional Sweden: aspects of tradition and transition, *Sociologia Ruralis*, 39(3), pp. 280–293.

EVANS, G. (2001) *Cultural Planning: an Urban Renaissance.* London: Routledge.

EVANS, G. (2003) Hard-branding the cultural city: from Prado to Prada, *International Journal of Urban and Regional Research*, 27(2), pp. 417–440.

FREY, B. S. (1994) The economics of music festivals, *Journal of Cultural Economics*, 18, pp. 29–39.

FRIDBERG, T. and KOCH-NIELSEN, I. (1997) *Cultural Capital of Europe, Copenhagen 96 (Konsekvensanalyse af Kulturby 96).* Copenhagen: Danish National Institute of Social Research.

GARCÍA, B. (2004) Urban regeneration, arts programming and major events: Glasgow 1990, Sydney 2000, Barcelona 2004, *International Journal of Cultural Policy*, 10(1), pp. 103–118.

GEERTZ, C. (1993) *The Interpretation of Cultures.* London: Fontana Press.

GGCVTB (2002) *Greater Glasgow and Clyde Valley Tourist Board, Facts and Figures: Value and Volume of Tourism* (www.seeglasgow.com).

GIBSON, L. and STEVENSON, D. (2004) Urban space and the uses of culture, *International Journal of Cultural Policy*, 10(1), pp. 1–4.

GOMEZ, M. V. (1998) Reflective images: the case of urban regeneration in Glasgow and Bilbao, *International Journal of Urban and Regional Research*, 22(1), pp. 1–16.

GREENWOOD, D. J. (1976) Tourism as an agent of change: a Spanish Basque case, *Annals of Tourism Research*, 3, pp. 128–142.

GREENWOOD, D. J. (1989) Culture by the pound: an anthropological perspective on tourism as cultural commoditization, in: V. SMITH (Ed.) *Hosts and Guests. The Anthropology of Tourism*, 2nd edn, pp. 171–185. Philadelphia, PA: University of Pennsylvania Press.

HALL, C. M. (1992) *Hallmark Tourist Events: Impacts, Management and Planning.* London: Belhaven Press.

HALL, P. (1988) *Cities of Tomorrow.* Oxford: Blackwell.

HANNERZ, V. (1988) *The world system of culture: the international flow of meaning and its local management.* (unpublished manuscript).

HANNIGAN, J. (2003) Symposium on branding, the entertainment economy and urban place building: introduction, *International Journal of Urban and Regional Research*, 27(2), pp. 352–360.

HARVIE, J. (2003) Cultural effects of the Edinburgh International Fesival: elitism, identities, industries, *Contemporary Theatre Review*, 13(4), pp. 12–26.

HEINRICH, N. (1988) The Pompidou Centre and its public: the limits of a utopian site, in: R. LUMLEY (Ed.) *The Museum Time Machine: Putting Cultures on Display*, pp. 199–212. London: Routledge.

HILLER, H. (1998) Assessing the impact of mega events: a linkage model, *Current Issues in Tourism*, 1(1), pp. 47–57.

HITCHCOCK, M. (1999) Tourism and ethnicity: situational perspectives, *International Journal of Tourism Research*, 1, pp. 17–32.

HOLND FSTVL (2002) *Future of Festival Formulae.* Amsterdam: Holland Festival.

HUGHES, G. (1999) Urban revitalization: the use of festival time strategies, *Leisure Studies*, 18(2), pp. 119–135.

HUGHES, H. (2000) *Arts, Entertainment and Tourism.* Oxford: Butterworth Heinemann.

ISAR, R. F. (1976) Culture and the arts festival of the twentieth century, *Cultures*, 3, pp. 125–145

JACKSON, P. (1988) Street life: the politics of carnival, *Environment and Planning D*, 6, pp. 213–227.

JAMIESON, K. (2004) Edinburgh: the festival gaze and its boundaries, *Space and Culture*, 7(1), pp. 64–75.

JUDD, D. (1999) Constructing the tourist bubble, in: D. JUDD and S. FAINSTEIN (Eds) *The Tourist City*, pp. 35–53. New Haven, CT: Yale University Press.

KLAIC, D., BOLLO, A. and BACCHELLA, U. (nd) *Festivals: challenges of growth, distinction, support base and internationalization*. Department of Culture, Tartu City Government, Estonia.

LANDRY, C., GREENE, L, MATARASSO, F. and BIANCHINI, F. (1996) *The Art of Regeneration: Urban Renewal through Cultural Activity*. Stroud: Comedia.

MARSTON. S. (1989) Public rituals and community power: St Patrick's parades in Lowell, Massachusetts 1841–1874, *Political Geography Quarterly*, 8(3), pp. 255–269.

McLAY, F. (1990) *The Reckoning: Public Loss, Private Gain (Beyond the Culture City Rip Off)*. Glasgow: Clydeside Press.

MUELLER, W. and FENTON, D. (1989) Psychological and community issues, in: G. SYME, B. SHAW, D. FENTON and W. MUELLER (Eds) *The Planning and Evaluation of Hallmark Events*, pp. 92–102. Aldershot: Avebury.

MUIR, E. (1997) *Ritual in Early Modern Europe*. Cambridge: Cambridge University Press.

MYSERCOUGH, J. (1991) *Monitoring Glasgow 1990*. Report prepared for Glasgow City Council, Strathclyde Regional Council and Scottish Enterprise.

NURSE, K. (1999) Globalization and Trinidad carnival: diaspora, hybridity and identity in global culture, *Cultural Studies*, 13(4), pp. 661–690.

OWUSU, K. and ROSS, J. (1988) *Behind the Masquerade: The Story of the Notting Hill Carnival*. London: Arts Media Group.

OZOUF, M. (1988) *Festivals and the French Revolution*. Cambridge, MA: Harvard University Press.

PADDISON, R. (1993) City marketing: image reconstruction and urban regeneration, *Urban Studies*, 30(2), pp. 339–350.

PRENTICE, R. and ANDERSEN, V. (2003) Festival as creative destination, *Annals of Tourism Research*, 30(1), pp. 7–30.

PUTNAM, R. D. (2001) *Bowling Alone: The Collapse and Revival of American Community*. New York: Simon & Schuster.

QUINN, B. (1998) *Authoring landscapes, shaping places: case studies of the Wexford Festival Opera (1951–1997) and the Galway Arts Festival (1978–1997)*. Unpublished PhD thesis, University College Dublin.

QUINN, B. (2003) Symbols, practices and myth-making: cultural perspectives on the Wexford Festival Opera, *Tourism Geographies*, 5(3), pp. 329–349.

QUINN, B. (2004) *Making space for festivity: Venetians' views on Carnevale*. Paper presented at the *IGU Commission on Tourism, Leisure and Global Change*, Pre-IGU Congress meeting, August, Drymen, Scotland.

QUINN, B. (2005) Changing festival places: insights from Galway, *Social and Cultural Geography*, 6(2) (forthcoming).

RICHARDS, G. (2000) The European cultural capital event: strategic weapon in the cultural arms race?, *Cultural Policy*, 6(2), pp. 159–181.

RICHARDS, G. and WILSON, J. (2004) The impact of cultural events on city image: Rotterdam, Cultural Capital of Europe 2001, *Urban Studies*, 41(10), pp. 1931–1951.

RITCHIE, J. R. (1984) Assessing the impact of hallmark events: conceptual and research issues, *Journal of Travel Research*, 23, pp. 2–11.

ROBERTSON, M. and WARDROP, K. (2004) Events and the destination dynamic: Edinburgh festivals, entrepreneurship and strategic marketing, in: I. YEOMAN, M. ROBERTSON, J. ALI-KIGHT ET AL. (Eds) *Festivals and Events Management: An International Arts and Culture Perspective*, pp. 115–129. Oxford: Elsevier Butterworth-Heinemann.

ROBINSON, M., PICARD, D. and LONG, P. (2004) Introduction: festival tourism: producing, translating an consuming expressions of culture(s), *Event Management*, 8, pp. 187–189.

ROCHE, B. (2004) Cork presents programme for its year as capital of culture, *Irish Times*, 8 October.

ROCHE, M. (1994) Mega events and urban policy, *Annals of Tourism Research*, 21, pp. 1–19.

RYAN, M. (2002) Culture win gave Glasgow boost, *BBC News Online* (http://news.bbc.co.uk).

SAMPATH, N. (1997) Mas' identity: tourism and local and global aspects of Trinidad carnival, in: S. ABRAM, J. WALDREN and D. V. L. MacLEOD (Eds) *Tourists and Tourism: Identifying with People and Places*, pp. 149–172. Oxford: Berg.

SCOTINFORM (1991) *Edinburgh Festivals study 1990/91: visitor survey and economic impact assessment*. Report to Scottish Tourist Board, Lothian and Edinburgh Enterprise Limited, City of Edinburgh District Council, and Lothian Regional Council. Edinburgh: Scotinform.

SCOTT, A. J. (2000) *The Cultural Economy of Cities*. London: Sage.

SMITH, M. P. (1991) *City, State and Market: The Political Economy of Urban Society*. Oxford: Blackwell.

SMITH, S. J. (1996) Bounding the Borders: claiming space and making place in rural Scotland,

Transactions, Institute of British Geographers, 18, pp. 291–308.

TURNER, V. (1982) *Celebration: Studies in Festivity and Ritual.* Washington, DC: Smithsonian Institution Press.

URRY, J. (1990) *The Tourist Gaze.* London: Sage.

WAITT, G. (2003) Social impacts of the Sydney Olympics, *Annals of Tourism Research,* 30(1), pp. 194–215.

WAITT, G. (2004) A critical examination of Sydney's 2000 Olympic Games, in: I. YEOMAN, M. ROBERTSON, J. ALI-KIGHT ET AL. (Eds) *Festivals and Events Management: An International Arts and Culture Perspective,* pp. 391–408. Oxford: Elsevier Butterworth-Heinemann.

WARD, S. (1998) *Selling Places: The Marketing and Selling of Towns and Cities 1850–2000.* London: Spon.

WATERMAN, S. (1998) Carnivals for elites? The cultural politics of music festivals, *Progress in Human Geography,* 22(1), pp. 54–75.

WILLEMS-BRAUN, B. (1994) Situating cultural politics: fringe festivals and the production of spaces of intersubjectivity, *Environment and Planning D,* 12, pp. 75–104.

ZUKIN, S. (1991) *Landscapes of Power: From Detroit to Disney World.* Berkeley, CA: University of California Press.

ZUKIN, S. (1998) Urban lifestyles: diversity and standardization in spaces of consumption, *Urban Studies,* 35(5/6), pp. 825–839.

The Global Cultural City? Spatial Imagineering and Politics in the (Multi)cultural Marketplaces of South-east Asia

Brenda S. A. Yeoh

Introduction

While cities have long been integral to the organisation of space beyond national boundaries, it is in the past two decades, with the efflorescence of the age of globalisation, that certain cities have become valorised as critical nodes and vital powerhouses of the global economy (Yeoh, 1999). Territorial reconfigurations of capitalism under conditions of economic globalisation have led to a 'rescaling' to sub-national scales, particularly cities (Bunnell and Coe, 2001). As Paul puts it, the global city

> now occupies a central analytical position in the literature on globalisation and the spatial organisation of global capitalism … In fact, attracting global fixed capital investment (corporate headquarters, production facilities, downtown skyscrapers) and circulating capital (transport, tourism, cultural events) through an international identity has become a nearly universal economic strategy (Paul, 2004, p. 572).

Such a strategy is by no means a purely economic one, as it is equally dependent on the manipulation of symbols and the construction of identities (Kong, 2000; Paul, 2004). As Clammer puts it

> It is the cultures of urban spaces that are most immediately and directly influenced by globalisation (in terms of consumption patterns and tastes, fashion, architecture, media and new forms of material culture) and equally it is urban cultures (intellectual trends, economic and technological innovations and again the media) which largely constitute so-called globalisation (Clammer, 2003, pp. 403–404).

The creation of global cities increasingly rests on the integration of economic and cultural activity around the production and consumption of the arts, architecture, fashion and design, media, food and entertainment, through what Ley (1996, pp. 9–10) calls a "croissants and opera" strategy, an upgraded version of "bread and circuses" intended for a more affluent and cosmopolitan clientele. This is accompanied by the emergence of a new mode of entrepreneurial governance in the city where the primary task of city officials is to ensure that the city becomes a hub of flexible production and consumption (Harvey, 1989), resulting in the intensification of interurban competition or "place-wars" (Silk, 2002, p. 777) for mobile investors, international talent and cosmopolitan élites. These 'place-wars' are indicative that "economic globalisation is conditioned upon political accommodations within and among nation-states" (Hill and Fujita, 2003, p. 209), putting paid to claims regarding the withering away of the nation-state and instead giving emphasis to the significant role of the nation-state in the politics of urban restructuring and changing spatial reorganisation in relation to globalisation. They are also no longer fought on purely economic grounds but across the diverse terrain of cultural policy including: the construction of prestigious urban flagship projects as part of the cultural regeneration of the city; production-based strategies, such as the development of a cultural industries sector; and, consumption-based strategies through image promoting and place marketing (Kong, 2000; Watkins and Herbert, 2003). No longer just epicentres of capital investments and transactions, global cities not only incorporate many of the cultural industries but are also sites of transnational cultural mixing and dynamic social foment (Yeoh, 1999). In addition, such cities provide epicentres from which symbolic flows emanate (Jansson, 2003).

This paper first reviews the discursive underpinnings of the growing vision of 'going global' adopted by many cities in south-east Asia as a strategy for urban growth and regeneration. It elaborates on the way economic and cultural policies have been integrated as a means of conserving 'local' heritage values and a sense of 'Asian' identity and, at the same time, advancing urban fortunes and futures amidst intense intercity, place-based competition. This is followed by an examination of the emerging spatial politics, social polarisations and questions of justice accompanying the increasingly significant impact of leisure and the aestheticisation of the urban landscape (Ley and Olds, 1988).

The Cultural Imagineering of the City: Visions and Mega-projects

Over recent decades in Asia, many cities have embraced an entrepreneurial regime and experienced dramatic transformations as they adapt to new, accompanying socioeconomic imperatives predicated on the 'mobilising myth' of becoming a global city. As a discursive category conjuring up imaginaries of high modernity, mega-development, 21st-century urbanity and progressive urban futures in the new millennium, "the 'global' has become an 'icon' or a spatial metaphor ... with considerable political power" (Kelly, 1997, p. 168). For example, as south-east Asian economies became increasingly integrated into regional and global circuits and rose to global prominence during the 1990s, their cities also began subscribing to the discourses and logics of globalisation, investing in the 'imagineering'[1] of global images as they jostle for a place in the new urban utopia.[2]

At the same time, given the colonial context from which many of these cities emerged just a few decades ago, the post-colonial enterprise of cultivating national identity and promoting national pride remains highly salient (Yeoh, 2001, p. 458). Indeed, in many parts of south-east Asia, while styles of governance may range from soft authoritarianism to more thorough-going statism, the cultural production of national identity remains a major project of the capitalist developmental state. As a key node in the post-colonial state, the city is often "overwhelmed with the onslaught of representational spaces" in attempts to produce the "ideal of the post-colonial citizen" (Srivastava, 1996, p. 406). Urban forms and architecture, in particular, have provided a means by which a post-colonial nation can construct a dialogue with its past (Kusno, 1998). While the post-colonial engagement with modernity in cities have not always been overimaginative—sometimes reduced to the question of how to be "Western without depending on the West"—cities wrestle with the need to find a balance between "cultural self-determination and international modernity" (Vale, 1992, p. 53).

Often, the need to signify global connections while promoting nationalist sensibilities takes concrete shape on the urban landscape in the form of prominent flagship projects. These 'urban mega-projects' (Olds, 1995) are constituted by large, high-profile and self-contained developments designed at least partly as a catalyst for urban regeneration (Beazley et al., 1996). This was especially the case in the euphoria prior to the Asian economic crisis of 1997–98 when the runaway economic success of Asian 'tigers' and 'dragons' emboldened regional élites to locate "the leading edge of global change" in Asia, reimagining "the [once] mystical, sleepy (post-)colonial Orient, as, at once, a new threat to Euro-American supremacy and a new paradise of economic opportunity" (Bunnell, 2004, pp. 5–6). Drawing on 'public–private partnership' as a key strategy in urban regeneration programmes in anticipating the dawn of the 'Asian century' as the old millennium closed, cityscapes were spectacularly transformed by the construction of mega-projects, including impressive waterfront cities, 'world-class' convention centres, festival marketplaces and cultural centres with "global urban-national visibility", "state-of-the-art" office complexes and other mixed-use commercial developments (MXDs) (Bunnell et al., 2002, p. 21). The accelerated creation of new urban forms (Forbes and Thrift, 1987) produced 'Asian mega-projects' such as the 'Golden Triangle' of Kuala Lumpur's city centre and the Petronas Twin Towers, Jakarta's Gateway Precinct, the new

commercial and cultural hub at Marina Centre in downtown Singapore, Tokyo's Teleport Town and Yokohama's Minato Mirai 21 project as well as the Luijiazui financial district in Shanghai's Pudong 'New Area' (Pow, 2002, quoting Olds, 1995, 1997, 1998; Cybriwsky, 1997, 1999; Cartier, 1998; Ford, 1998; Bunnell, 1999).[3] As Olds (1995) points out, these high-profile mega-projects constitute both a real and a symbolic dimension aimed at linking the city to the global economy and are dependent on a host of factors closely associated with the globalisation of finance and property markets, the growth of transnational corporations as well as the creation of social networks between policy-makers and mobile urban élites and professionals. At the same time, these monumental consumerist spectacles (Harvey, 1993, p. 24) are often "abstracted from local culture and translated as symbols of *the* culture to be promoted beyond a nation's own borders" (Firat, quoted in Silk, 2002, p. 779; original emphasis). In the rest of this section, I examine some examples of the "meta-narratives of place" (Chang and Lim, 2004, p. 167)—often woven around significant nodes such as mega-projects—constructed by cities in Asia in their attempt to connect to a global imaginary, while simultaneously appropriating the 'cultural' realm as a means of maintaining a sense of unique identity.

Cultural imagineering is sometimes a strategy of reconstruction occasioned by regime change. In post-1998 Jakarta,[4] amidst an "unruly urban milieu" and the physical destruction of the old urban core, the urban élite were anxious to construct a new narrative of "nationalist urbanism" as a means to recreate order and authority in order to restore the nation's image to tourists and the world outside as well as reimagine a "coherent image of the past to regain power and influence" (Kusno, 2004, p. 2389). This was predicated on producing a 'nostalgia for orderliness', through establishing projects of cultural preservation and urban beautification such as the restoration of Batavia, the colonial town, the renovation of the water fountain of the Hotel Indonesia (HI) traffic island (featuring the Welcome Statue built in the 1960s under the instructions of President Sukarno to welcome Asian Games athletes to the city), the fencing of National Monument (Monas) Park and the erection of statues of national heroes at street junctions (Kusno, 2004). By drawing on and regenerating urban remnants of the past, Jakarta's governing élite attempted to project the capital as a city of centrifugal power, a task of increasing urgency even as city life unravels amidst social, economic and political crises in the post-Suharto state.

Elsewhere, where more stable socio-political conditions have prevailed, the cultural imagineering of the city is often motivated by competition to gain a foothold higher up the global cities league table. Singapore is the quintessential entrepreneurial city, where the relentless drive to establish the city-state as an economic powerhouse and international business hub through promoting and attracting high-growth investments, financial services and value-added services has been the main preoccupation for the past four decades (Pow, 2002, p. 154). While developing the city-state as an international base for the headquarters of transnational corporations in manufacturing and services was the predominant priority up to the 1980s, the 1990s saw a diversification of globalisation strategies[5] to include developing the arts, culture and entertainment in the bid to become a 'world-class city'. This is a strategy already put to work in European cities such as Glasgow, Athens, Brussels and Amsterdam, where concepts such as the 'arts city', 'city of culture' and 'cultural capital' have been deployed to attract tourists and capital investments, create new urban imaginaries, provide opportunities for employment and reuse defunct urban zones and buildings, whilst encouraging residents to rediscover the city centre (Bianchini, quoted in Chang, 2000a, p. 820; see also Bassett, 1993; Griffiths, 1995; Hubbard, 1996; Crewe and Beaverstock, 1998; McCarthy, 1998; Waterman, 1998; Hall and Robertson, 2001; Teedon, 2001; Gotham, 2002; Watkins and Herbert, 2003; Bayliss, 2004).

Singapore's move into the cultural arena in the 1990s was not entirely unanticipated. It was heralded by a major shift of urban policy in the second half of the 1980s away from the earlier 'demolish-and-rebuild' philosophy responsible for creating a clean, orderly if somewhat antiseptic environment towards a revalorisation of the older urban fabric such as Chinatown, Little India and Kampong Glam (the Malay quarter) as 'historical districts' or 'heritage areas' (Yeoh and Kong, 1994; Yeoh and Huang, 1996; Chang and Yeoh, 1999; Chang, 2000b). This policy turn was partly driven by the state's interest in maintaining local cultural heritage as a bulwark against what was perceived then as the rapid infiltration of Western values as Singapore modernised. At that time, many among the governing élite were wary that, while 'Westernisation' had served Singapore well in its quest for industrialisation and economic development, the city-state was also in danger of losing its 'Asian' roots and identity. As Goh Chok Tong, then First Deputy Prime Minister, declared in 1988

We are part of a long Asian civilisation and we should be proud of it. We should not be assimilated by the West and become a pseudo-Western society. We should be a nation that is uniquely multiracial and Asian, with each community proud of its traditional culture and heritage (Goh, 1988, p. 15).

The shift towards heritage conservation and creation in the city in the mid 1980s was also economically motivated by the need to reclaim the city's oriental mystique and charm in order to arrest falling tourist numbers (Yeoh, 2000, p. 118). As Chang et al. observe

Heritage conservation constituted one element of multi-faceted redevelopment strategies designed to cater to tourist demands for uniqueness on the one hand, while providing an opportunity to improve urban aesthetics on the other (Chang et al., 1996, p. 294).

What is already evident during this phase is the discursive construction of heritage conservation and aesthetic considerations as desirable urban innovations compatible with developmentalist goals and economic pursuits (although conservation–redevelopment tensions continue to be very real concerns at the level of policy implementation).

By the 1990s, state discourses around the multiply inflected, but thought-to-be congruent, relationships between 'heritage, culture and aesthetics', 'local identity', 'tourism', 'economic development', 'global reach' and the 'cosmopolitan city' had shifted to a higher gear in Singapore. In 1992, the Singapore government coined the term "Global City for the Arts" to spearhead its vision of cultivating a thriving arts, cultural and entertainment scene, both to further sociocultural objectives of enriching the local cultural scene and fostering national pride, as well as a socioeconomic strategy not only to attract tourists, but also to compete for, welcome and retain "foreign talent"—transnational élites of the entrepreneurial, managerial and professional class who "will add to Singapore's vibrancy and secure our place in a global network of cities of excellence" (Lee Kuan Yew, then Senior Minister, quoted in Yeoh, 2004, p. 2435). In this vision, the arts was expected to contribute to creating a 'symbolic economy' with three nodes: an art and antique trading and auction centre; a theatre hub of south-east Asia; and, an entertainment destination for tourists and leisure-seekers (Chang, 2000a, p. 819).

By the turn of the millennium, the 'Global City for the Arts' vision was further complemented by the notion of becoming a "Renaissance City", a term introduced by the then Prime Minister, Goh Chok Tong in his 1999 National Day Rally, and elaborated upon in the state's "Renaissance City Report" released in March 2000. As Lee Yock Suan (2001, p. 2), the then Minister of Information and the Arts explained, the state drew on the term "renaissance"—taken to mean "the spirit of creativity, innovation and multidisciplinary learning and of socioeconomic, intellectual and cultural vibrancy"—to articulate a vision for Singapore in the new knowledge-based economy and to explain "how culture and the arts can contribute to the national picture" as well as "project a positive and well-rounded image of Singapore internationally". As with the earlier discourses around the insertion of heritage into the city, the twin aims of nurturing Singaporeans "with a deep appreciation for the arts and keen sense of aesthetics" and developing and supporting "both local and overseas creative talents" emphasised the need to appeal to international talent as well as enhance "our Asian heritage … even as we evolve a Singaporean identity" (Lee Yock Suan, 2001, p. 2).

Central to this vision is the monumental urbanscape launched in October 2002 and costing an estimated S\$667 million, The Esplanade-Theatres on the Bay,[6] an assemblage of several theatres and performance spaces of spectacular shape and aesthetic design covering 6 hectares of prime land that visibly transformed the city's skyline, waterfront and aerial view. While state discourse made clear that the Esplanade is a major economic investment expected to generate returns by putting Singapore on the map for arts and cultural tourism, it also envisaged a nation-building role for such a landscape of spectacularity, not least by recalling the 'Asian' roots of Singapore arts. The then Prime Minister Goh Chok Tong (quoted in Kong and Yeoh, 2003, p. 189)—for example, articulated the view that the Esplanade should "evoke in Singaporeans an appreciation and understanding of the old civilisations they all belonged to, and the part of the world they were in", especially "as Singapore developed and became more open to the external cultures and influences, and as Singaporeans traveled and absorbed different values". As a 'world-class' performing arts centre, the Esplanade is built to "usher in a new Asian Renaissance" (*The Straits Times*, 10 November 1994).

While culture-led urban regeneration in a climate of intensified place-based competition is most manifestly illustrated in the case of the arts and heritage industries (as shown above), the appropriation of the cultural sphere also takes place in other ways, as part of an urban cosmetics or aesthetics policy, or in the form of cultural markers to assert local differentiation and at the same time reinforce corporate power and entrepreneurial schemes. For example, place-marketing strategies promoting the S$2 billion flagship project Suntec City—comprising a mega-size convention and exhibition centre, integrated with five 'state-of-the-art' office towers and a giant retail mall and entertainment centre—are elaborated through an appropriation of 'exotic' discourses on Asian symbolism and Chinese geomancy (Pow, 2001, p. 118). Taking pride of place in Singapore's new 'commercial and cultural hub' at Marina Centre, Suntec City is said to combine "powerful Asian elements such as the Hindu *mandala* and Chinese ideas of *feng shui* and *yin-yang*" to provide a geomantically auspicious entrepreneurial landscape where the notion of an "open-handed exchange of prosperity" is captured by the architectural design of placing a gigantic "Fountain of Wealth" (symbolising a gold ring) in the middle of a palm (represented by the four office towers (the four fingers) and the 18-storey office block (the thumb)) (*Suntec City Commemorative Book*, quoted in Pow, 2001, p. 118). For both the Hong Kong tycoon investors and ordinary visitors, the inwardly cascading, constantly flowing waters of the fountain conjure up a vision of "riches pouring in", imbuing speculative real-estate development as well as a visit to the mall with an aura of "divine providence". By consciously manipulating architectural spectacle and appropriating cultural 'myth', entrepreneurial landscapes such as Suntec City serve as a means of reproducing and legitimising global capitalist power and interests (Pow, 2001).

Turning to another example of an entrepreneurial (as opposed to an overtly 'cultural' arts or heritage) landscape, Bunnell (2004) argues that the Kuala Lumpur City Centre (KLCC) mega-project (which includes the Petronas Towers, the world's tallest building when it was built) could be read as a form of state-sponsored architectural nationalism, performing

a symbolic civic role both internationally and domestically: marking the city and nation on global maps as a modern and 'investible' metropolis, and demonstrating that *Malaysia boleh*, a 'can do' attitude free from the supposed shackles of (neo-) colonial inferiority (Bunnell, 2004, p. 9).

While earlier attempts at the height of the National Cultural Policy of the 1980s to deploy architecture and urban design as signs of national transformation had drawn on supposedly recognisably 'Malaysian' symbols and styles—such as a roof shaped in the form of a Malay *keris* or dagger in the case of the Menara Maybank building and the five columns representative of the five pillars of Islam in the case of the 'LUTH' Tower, both in Kuala Lumpur—the Petronas Towers (comprising twin 84-storey towers made of stainless steel clad in glass and spanned at the 41st floor by a sky bridge) primarily distinguished itself by its record-breaking height as a "cultural landmark" (a phrase used by then Prime Minister Mahathir Mohamad, quoted in Bunnell, 2004, p. 70) symbolising the nation's world-class prestige, international visibility and upward mobility (Bunnell, 2004, p. 72). Instead of simply tacking on ostensibly cultural motifs, the Petronas Towers, like Singapore's Suntec City (promoted as the Asia–Pacific region's largest convention centre and the site of the 'world's largest fountain' as recorded in the 1998 Guinness Book of Records), engaged well-known international design and planning teams in order to create prestigious and high-impact signature structures (Ho, 2002) and which linked the "cultural value of architect[ure] with the economic value of land and building" (Zukin, 1991, p. 45). The 'climactic' effects of the Towers' superlative height were also multiplied many times over to reach international audiences through Fox Films, with whom the Malaysian government negotiated to showcase the Towers in Fox's 1999 romantic thriller, *Entrapment* (Silk, 2002, p. 784). Said the film's director, Jon Amiel

After audiences watch Mac [Sean Connery] and Gin [Catherine Zeta Jones] flee their pursuers by swinging underneath the sky bridge, 750 feet above Kuala Lumpur, I expect the towers could very well become a new cinematic landmark (Amiel, quoted in Silk, 2002, p. 784).

By eschewing ethnically exclusive architectural references[7] and embracing culturally innocuous superlatives, the Petronas Towers are intended to signify the new Malay in an increasingly multicultural society engaging the world of commerce (Kusno, cited in Bunnell, 2004, p. 75).

Other entrepreneurial landscapes ranging from whole new cities to sports complexes to host hallmark events and university campuses may also be read as instances of cultural production intended to fulfil the dual function of fuelling the nation's globalising ambitions on the one hand and producing new post-colonial citizens and shaping local sensibilities on the other. For example, while the Siliconising of the landscape through the creation of the 'intelligent cities' of Putrajaya and Cyberjaya along Kuala Lumpur's Multimedia Super Corridor is part of a broader symbolic intervention to leave a mark on global urban imaginative geographies, it is also predicated on culturally inflected meta-narratives about progressive 'Asian values' and the production of 'intelligent citizens' capable of innovation, self-learning and navigating the Information Age (Bunnell, 2004; Chang and Lim, 2004). In similar fashion, the 1998 Commonwealth Games held in Kuala Lumpur reframed and represented cultural symbols and narratives (through ceremonies, choreography and even in the choice of colours in the Games Signature) in order to press Malaysia's competitive advantage in the global marketplace and claims to world leadership as well as to signify Malaysia's aspirations to Vision 2020—Prime Minister Mahathir's plan to re-engineer the social, political and economic climate of Malaysia to achieve balanced economic growth, a high quality of life and the creation of *one* Malaysian race (*Bangsa Malaysia*) (Silk, 2002).

Turning attention to Indonesian attempts to produce the most appropriate architectural forms through a selective retrieval of indigenous and colonial cultures, Kusno's (1998) study of the architectural design of university campuses in Bandung and Jakarta illustrates the fundamental split inherent in post-colonial societies between a denial and displacement of 'colonial origins' on the one hand, and a recitation of the coloniality of 'Indonesian architecture' on the other, two strains which are perpetually contradictory and yet indissolubly intertwined. In sum, the cultural imagineering of the city in south-east Asia is both a growing enterprise intimately connected to commodity production and consumerism as well as a fraught terrain, requiring careful negotiation between global modernity, post-colonial sensibilities and nationalist dreams of a utopian future.

Cultural Politics and Contested Spaces

Current efforts put into urban image-making and branding are both self-generating and somehow peculiar. Since every new market-message is contested by the pluralism of urban social life ... there can never be a final, intersubjectively shared city image. Rather, ... the more contradiction and negotiation there are, the more resources may be put into image-making. And, the more effort that is put into the diffusion of a dominant image, the more image-creation must actually overlook the authentic complexities of social life (Jansson, 2003, p. 478).

Inasmuch as the hegemonic scripts underpinning visionary urban discourses and spectacular mega-projects are legitimised by appropriations of the 'cultural' sphere, they may also be ruptured. As Jansson puts it

The creation of the city image is not only a matter of cultural policy. What the city actually becomes, and how different groups experience it, depend on the activities of social actors as well as systemic forces ... [such as] alternative and oppositional groupings ... challenging such points of view (Jansson, 2003, p. 464).

While the cultural imagineering of the city often attempts to mask social, ethnic, class and gender polarisations by "mobiliz[ing] every aesthetic power of illusion and image" (Silk, 2002, p. 778), it is by no means an uncontested process (and in fact is a dynamic, self-generating one, as Jansson (as quoted above) indicates).

In south-east Asian urban life, the dynamics of such cultural politics are highly complex. While many urban civil society institutions are mainly products of the middle classes, there are also a burgeoning number of urban social movements—religious, ideological, cultural, ethnic, gendered and those emerging from the experience of the underclass—which propel competing visions of urban life that are opposed to, or seek to transcend, civil society as defined by the middle classes (Clammer, 2003). Seldom does the groundswell of competing visions and demands coalesce into a unified collective response (see later); instead, the contradictory qualities of imagineering the city to satisfy the twin goals of affirming national identity and extending global reach often lead to "integration, fragmentation, polarisation and reterritorialisation of numerous superimposed social spaces" (Lefebvre, quoted in Clammer, 2003, p. 416). As cities re-evaluate the nature of their relationship between the local and the global they set in motion a simultaneous politics of "forgetting" and "remembering", of "inclusion", "exclusion" and "revalorisation" (Lee and Yeoh, 2004, p. 2298)

At the discursive level, despite (and perhaps because of) the hyperbole and rhetoric in which all-encompassing vision statements (so central to cultural imagineering efforts) are wrapped, they are often not immune to counter- or mis-representations, sometimes in ways akin to parody. For example, in view of the Singapore government's vision to produce Singapore as the 'Global City for the Arts', critics have coined the term 'Global City for the Borrowed Arts' to capture concerns that the vision privileges high-cost mega-structures and commercially driven blockbuster events such as the staging of Broadway shows and foreign pop concerts to the detriment of smaller, local and non-profit productions and "Singapore experimental art" (Chang, 2000a, p. 826). Reflecting the views of many local artistic practitioners, one of the most respected of local playwrights, Kuo Pao Kun asked pointedly, "Can we have a Singapore Arts Centre by just bringing all the arts of the world to Singapore without our own education, without our own creativity?" (quoted in Kong and Yeoh, 2003, p. 183). These views were in turn countered by the state as a form of "false nationalism", a form of "protectionism" antithetical to the "national spirit" which is a "cosmopolitan" one (George Yeo, then Minister of Information and the Arts, quoted in Kong and Yeoh, 2003, p. 184).

At the level of specific mega-projects, meanings and symbolisms conferred on entrepreneurial landscapes by state or corporate powers are not always hegemonic but constantly inflected, unsettled and challenged by the possibility of alternative readings on the part of others (consumers and social groups with different interpretations and claims on the landscapes). A sense of parody is again involved in the rechristening of the Esplanade-Theatres on the Bay among locals who preferred to refer to the twin domes as "durians", "soursops", "porcupines" or the "gigantic housefly's eyes", as well as in the redesigning of the Petronas Towers as the "twin dipsticks" for air pollution levels in a city where residents apparently put less faith in officially released meteorological figures (particularly during heavy haze) than in calibrations based on visibility tests trained on the Twin Towers.[8]

The constant reworking of meaning is also present in a number of projects which combine 'heritage' and 'enterprise' in the production of 'new' urban spaces in post-colonial nations, often leading to a contest of meanings and priorities around questions such as what constitutes 'history' in a multiethnic post-colonial context, what should be valorised as 'heritage' (and hence what should be excised as of no 'heritage' value), who should control (and benefit from) the whole process of transforming the landscape and for what purpose (such as nationalism and tourism) (Cartier, 1993; Parenteau *et al.*, 1995; Yeoh and Huang, 1996; Chang, 2000b; Teo, 2003). For example, the designation of Singapore's Kampong Glam Historical District as a bounded area enjoying conservation status sparked controversial discussions in some quarters among the Malay/Muslim community as to what constitutes 'Malay heritage and culture'. Particularly contentious was the eviction of the descendants of Sultan Hussein Shah, the 19th-century pre-colonial ruler of Johor and Singapore, from their 'ancestral home' at the *istana* (palace) in the heart of Kampong Glam to make way for a S$16.7 million state-driven restoration project to convert the Istana Kampong Glam into a Malay Heritage Centre (opened in 2004) to showcase "the history, traditions, culture and future challenges of the Malay community" (*The Straits Times*, 16 August 2004), for this pits the notion of 'heritage' residing in the form of bounded 'heritage objects' isolated and displayed in a museum against the claims of 'family genealogy' and 'royal descent'. Not only does the objectification of heritage artifacts reify a particular version of heritage, the legal codification and subsequent naturalisation of the entire Historical District further serve to demarcate what is and is not heritage, for immediately outside the boundary line delimiting the conservation area, landscapes reflecting Malay culture such as a *madrasah* (religious school) and Muslim cemetery land face the threat of erasure. Conflicting discourses as to what constitutes 'heritage' are hence often generated by urban conservation projects that

> slice up the organic form and texture of cultural hearths in an arbitrary fashion, legislating boundaries between a defended zone perceived to be of historical value and an excluded landscape which is threatened with excision (Yeoh and Huang, 1996, p.421).

Competing versions of cultural authenticity also crop up elsewhere. For example, in order to illustrate how people's actual experiences of flagship projects might diverge from the boosterism associated with property development and place entrepreneurialism. Pow (2001, p. 131) identifies a spectrum of "*differentiated* and often *fragmented* consumption of place images and meanings" held by different social groups and users of Singapore's Suntec City. These range from: geomancers who came up with a rash of counter-representations and disagreements over the 'authenticity' and 'accuracy' of Suntec City's *feng shui* symbolism;[9] to office managers who were more concerned with practical issues such as the rental cost of offices, accessible location and the quality of floor spaces and aloof to what they saw as marketing gimmicks; to shoppers, users of the space and other observers who found the design of the mega-complex "daunting" and "alienating", likening the "Fountain of Wealth" to a "gigantic piece of kitsch" and the enclosed mega-spaces to an "urban fortress" which turns inwards on itself, "trapping" people in the buildings and emptying the surrounding streets of meaningful social activities (Pow, 2001). What emerges from a variety of urban encounters is a "kitchen of meanings" (Barthes, quoted in Pow, 2001, p. 132), where different people draw upon their subjective experiences and personalised 'expertise' to engage in the "hyper-imaginative" rescripting of Suntec City's symbolic landscape. As in the case of Suntec City, the "symbolic discontent" and "popular reworking" of meanings around Kuala Lumpur's Petronas Towers cannot be expected to have any "coherent or unified authoritative intentionality" (Bunnell, 2004, p. 78). Instead, symbolic reworking was to be found amidst everyday rumours, ranging from who was *behind* the project (in particular, querying the role of the oil company Petronas) and what was *beneath* the project (the towers having been moved 50 metres from their original location on account of limestone caverns close to the surface), to more politically charged suspicions that the 'intelligent' building is but a high-tech instrument of state surveillance (Bunnell, 2004, pp. 80–83).

Beyond the level of discursive negotiations and the circulation of counter-images, the cultural politics accompanying the imagineering of the city also animate the social materiality of the city in the form of social conflicts and urban encounters, framed by exclusion, sidelining, erasure, loss, fragmentary places and "out-of-place others" on the one hand, and the "emergent 'counter-global' spatialities of power" (Bunnell and Nah, 2004, p. 2447) seeking to unmake the processes of marginalisation, on the other. As Lees (2001) notes, the cultural politics of place entail an engagement with embodied, everyday social practices through which urban spaces are actively used and appropriated. As the pre-eminent space of encounters with difference, the city is perpetually poised at a moment of transformation—this is true of cities witnessing unprecedented political change, such as Jakarta where intense rupture has opened up spaces for "a more responsible, diverse and democratic urbanism" as well as "the formation of a new urban politics that is more egotistic, deliberately unpredictable and violent" (Kusno, 2004, p. 2391), as it is also true of cities characterised by relentless economic makeovers driven by the globalising ambitions of strong states.

Given the pluralism of urban social life, reactions to the cultural imagineering of urban spaces and the construction of urban mega-projects are seldom unified; instead, they often entail alliances, negotiations and conflicts among multiple actors with different interests and subjectivities. On the rare occasion, subterranean tensions between different groups may culminate into flashpoints, as occurred during the 1998 Commonwealth Games at Kuala Lumpur. Two days prior to the Games' closing ceremony, political, ethnic and economic tensions surfaced in the city, ignited by the public sacking and humiliation of the then deputy prime minister Ibrahim Anwar. Anwar's dismissal for the alleged sodomy of two junior government ministers acted as a catalyst for anti-government demonstrators who gathered on the streets and in Merdeka (Independence) Square where they were caught on film by a number of foreign news crews that had been allowed into Malaysia to cover the Games (Silk, 2002, pp. 789–790).[10]

More commonly, collective action coalescing around urban projects goes beyond street-level confrontations. For example, Bunnell and Nah (2004) examine the way subordinated groups—in this case, squatters and the indigenous Orang Asli minority who have been displaced as a result of the global reorientation process of greater Kuala Lumpur—actively articulated and demonstrated in-place identities by drawing on a wide repertoire of transnationally networked resources to assert their land rights within and beyond Malaysian courts. By mobilising international declarations of housing and human rights and connecting up with other indigenous peoples beyond the nation-state, these groups were able to strengthen their claims to remaining 'in-place' and to counter the exclusionary practices of the state's globalising vision.

Also focusing on Malaysia but further north, Teo (2003) points to the limits reached in imagineering Penang as the 'Silicon Valley of the East' as well as a major tourism destination offering sun, sea, heritage and culture. Increasing disillusionment among the people of Penang, who felt peripheralised by what they saw as the reorientation of the city towards tourists at the expense of locals, reached a climax with the repeal of the Rent Control Act in 2000 to make way for more commercially viable buildings. Contestory strategies on the ground were bifurcated: while the Penang Heritage Trust, an élite non-government organisation, advocated the conservation of the urban fabric and the conversion of old buildings into restaurants, boutique hotels and art galleries, the residents of Penang's Chinatown and Little India—many of whom are illiterate, working-class or self-employed tenants living in rent-control shophouses—organised themselves into a tenants' self-help group called Save Ourselves in order to appeal against eviction orders and protest against the gentrification of traditional areas. The physical appropriation of the streets as a strategy of reclamation was also drawn upon on the eve of the New Year of 2001 when 500 tenants gathered for a candlelight vigil at Noordin Street "as a display of solidarity in their quest to save old Penang from the bulldozers and the yuppies" (Teo, 2003, p. 559).

The terrain of cultural politics around urban mega-projects may also be traversed by an unorchestrated array of unintended users who trace everyday geographies in counterpoint to the grand narratives and visions spun by the producers of such projects. Pow (2001)—for example, shows how corporate spaces in Suntec City are frequently appropriated by all manner of unexpected consumers: students appropriating air-conditioned places for studying, or hanging out; manual workers retreating to the benches and shelters for afternoon naps; gays frequenting the toilets as 'cruising' spots; and myriad others whose ordinary routines become acts of transgression not in themselves, but as a result of incongruous juxtaposition with the intended spectacularity of these spaces. As Pow (2001, p. 150, quoting Raban) puts it, "concealed beneath the material reality and dominant representation of Suntec City, there exists a plethora of interpenetrating 'soft cities' made hard through tactics of spatial occupation and appropriation".

(Multi)cultural Marketplaces and Cultural Justice

In Hannerz's (1993) interpretation of the classic theme of the cultural role of world cities, he focuses attention on world cities as cultural marketplaces, places of 'global cultural brokerage' involving a high density of local as well as transnational relationships between highly skilled professional and business élites; low-skilled, low-income migrant workers; specialists in expressive activities; and world tourists. As a site continually renewed by the "creative bloodstream" (Hall, 2000, p. 646) of transnational flows of different peoples, the contemporary global (ising) city is an "absorptive, continuously changing terrain that incorporates new cultural elements whenever it can" (Hall, quoted in Sassen, 2000, p. 176). Yet, as Sassen (2000) argues, such an observation needs qualification, because contemporary

global culture is in a constant state of flux and contestation which is in turn characterised by diverse work cultures and cultural environments in which the cultural politics at work are often capable of transforming the centre.

It is precisely the (multi)cultural wealth created by the intensity of transnational urban encounters in the city—what Jacobs (1996, p. 4) calls "the very place of our meeting with the other"—that makes the city a profitable place where the governing élite and corporate power could capitalise on immense cultural energy as a basis for economic regeneration. As this paper has shown, as in Europe and North America, culture-led urban regeneration has indeed been a major force of change in many south-east Asian cities. Unlike the Western developed world context, however, the use of cultural imagineering, urban mega-projects and iconic architecture as an urban regeneration strategy tends to be more spatially concentrated in the capital and large regional cities, further widening the gap between the mega-cities intent on extending their global reach at the pinnacle of the urban hierarchy and the 'fourth world cities' (Shatkin, 1998) at the bottom of the hierarchy which are perceived to be structurally irrelevant to the current round of global capital accumulation.[11]

It is hence in the already-large, globalising cities of south-east Asia that governments have invested considerable resources to take the global imaginary to further heights. At the same time, it is also the cultural thickness of urban encounters that underscores the vitality of cultural politics at work in the globalising city, both at the level of counter-representations and alternative spatial discourses, as well as in terms of collective social action and everyday practices, as has been richly articulated in current urban scholarship on south-east Asian cities. While questions of civility and inclusive notions of citizenship remain fraught terrains in deeply plural, post-colonial societies characterised by authoritarian governments and poorly developed civil space, urban imagineering as a form of cultural globalisation has brought to the fore not only new urban discourses but also oppositional tactics which chisel away at the image and edifice of 'the global cultural city'. These may not produce counter-hegemonic visions powerful enough to challenge state-led plans and programmes and reshape urban space, but allow the inhabitants of the city to continue to stake multiple claims to the city. Yet, what remains at large—and therefore where scholarship must surely turn its attention soon—is the broad question of "cultural justice" (Morrison, 2003, p. 1629), involving a fundamental revaluing of diverse peoples—not only across divisions of race, class, gender, religion and language, but also across nationality and citizenship lines—encompassing both cultural and economic dimensions in order to achieve greater inclusion of cultural difference—which is, after all, the hallmark of a global city.

Notes

1. According to Paul (2004, p. 574), the verb 'to imagineer' was coined by Walt Disney Studios to describe its strategy of "combining imagination with engineering to create the reality of dreams" in creating theme parks.
2. This does not detract from the fact that many primate cities in the region continue in the grip of major urban social and economic problems including fast-growing populations, rapid in-migration, sub-standard housing and infrastructure, high rates of unemployment and underemployment, ethnic and religious conflict, widening socioeconomic inequalities and degraded environments (Clammer, 2003, p. 406).
3. The development of a 'cultural policy' or a 'cultural industry' as part of civic boosterism and entrepreneurship is most acute in, but not confined to, capital cities with globalising ambitions. It is also an important strategy in the smaller cities (for example, see Yang and Hsing, 2001, on Kinmen; Teo, 2003, on Penang).
4. The riots of May 1998 in Jakarta were instrumental to the collapse of Suharto's New Order government.
5. Other plans in the 1990s included establishing Singapore as an e-commerce hub for hi-tech companies, an educational centre for international institutions, a regional medical centre and a 'tourism capital' (Chang and Yeoh, 1999; Olds and Yeung, 2004).
6. The Esplanade-Theatres on the Bay originated from a recommendation made by the Advisory Council on Culture and the Arts, appointed in 1988 to review and chart a new direction for Singapore's arts development (Kong and Yeoh, 2003, p.178)
7. Cultural references were not entirely absent inside the Towers. Apparently, the floor plans are based on a motif of interlinking squares and circles representing an ancient Muslim symbol of harmony and strength, and decorated with Malaysian wood and stone (Silk, 2002, p. 784).
8. Also in popular circulation is the comparison of the Twin Towers to upturned *jagung* (corn cobs), in imminent danger of collapse if not for the sheer force of the then Prime Minister Mahathir's personality propping them up (Bunnell, 2004, p. 81).
9. Pow (2001, p. 136) notes—for example, that most of the geomancers referred to the "Fountain of Wealth" as an "egotistical" and "self-serving" monument of the Hong Kong tycoons rather than a symbol of the benevolence of the global property developers as intended by the tycoons.
10. Also see Waitt (1999) on the way another hallmark sporting event, the Sydney Olympic Games, was deployed as a medium of political discontent.
11. This does not mean that these cities are not impacted by global restructuring or integrated in some ways into the world system (see Lee and Yeoh's (2004) *Urban Studies* Special Issue on "Globalisation and the Politics of Forgetting").

References

BASSETT, K. (1993) Urban cultural strategies and urban regeneration: a case study and critique, *Environment and Planning A*, 25, pp. 1773–1788.

BAYLISS, D. (2004) Creative planning in Ireland: the role of culture-led development in Irish planning, *European Planning Studies*, 12, pp. 497–515.

BEAZLEY, M., LOFTMAN, P. and NEVIN, B. (1996) Downtown redevelopment and community resistance: an international perspective, in: N. JEWSON and S. MACGREGOR (Eds) *Transforming Cities: Contested Governance and New Spatial Divisions*, pp. 181–192. London: Routledge.

BUNNELL, T. (1999) Views from above and below: the Petronas Twin Towers and/in contesting visions of development in contemporary Malaysia, *Singapore Journal of Tropical Geography*, 20, pp. 1–23.

BUNNELL, T. (2004) *Malaysia, Modernity and the Multimedia Super Corridor: A Critical Geography of Intelligent Landscapes*. London: RoutledgeCurzon.

BUNNELL, T. and COE, N. (2001) Spaces and scales of innovation, *Progress in Human Geography*, 25, pp. 569–589.

BUNNELL, T. and NAH, A. M. (2004) Counter-global cases for place: contesting displacement in globalising Kuala Lumpur metropolitan area, *Urban Studies*, 12, pp. 2447–2467.

BUNNELL, T., DRUMMOND, L. and HO, K. C. (Eds) (2002) *Critical Reflections on Cities in Southeast Asia*. Singapore: Times Academic Press.

CARTIER, C. (1993) Creating historic open space in Melaka, *Geographical Review*, 83, pp. 359–373.

CARTIER, C. (1998) Megadevelopment in Malaysia: from heritage landscapes to 'leisurescapes' in Melaka's tourism sector, *Singapore Journal of Tropical Geography*, 19, pp. 151–176.

CHANG, T. C. (2000a) Renaissance revisited: Singapore as a 'Global City for the Arts', *International Journal of Urban and Regional Research*, 24, pp. 818–831.

CHANG, T. C. (2000b) Singapore's Little India: a tourist attraction as a contested landscape, *Urban Studies*, 37, pp. 343–366.

CHANG, T. C. and LIM, S. Y. (2004) Geographical imaginations of 'New Asia–Singapore', *Geografiska Annaler*, 86B, pp. 165–185.

CHANG, T. C. and YEOH, B. S. A. (1999) 'New Asia-Singapore': communicating local cultures through global tourism, *Geoforum*, 30, pp. 101–115.

CHANG, T. C., MILNE, S., FALLON, D. and POHLMANN, C. (1996) Urban heritage tourism: the global–local nexus, *Annals of Tourism Research*, 23, pp. 284–305.

CLAMMER, J. (2003) Globalisation, class, consumption and civil society in south-east Asian cities, *Urban Studies*, 40, pp. 403–419.

CREWE, L. and BEAVERSTOCK, J. (1998) Fashioning the city: cultures of consumption in contemporary urban spaces, *Geoforum*, 29, pp. 287–308.

CYBRIWSKY, R. (1997) From castle town to Manhattan town with suburbs: a geographical account of Tokyo's changing landmarks and symbolic landscapes, in: P. P. KARAN and K. STAPLETON (Eds) *The Japanese City*, pp. 56–78. Lexington, KY: University Press of Kentucky.

CYBRIWSKY, R. (1999) Changing patterns of urban public space: observations and assessments from the Tokyo and New York metropolitan areas, *Cities*, 16, pp. 223–231.

FORBES, D. and THRIFT, N. (1987) International impacts on the urbanisation process in the Asian region: a review, in: R. FUCHS, G. JONES and E. PERNIA (Eds) *Urbanisation and Urban Policy in Pacific Asia*, pp. 67–87. Boulder, CO: Westview Press.

FORD, L. R. (1998) Midtowns, megastructures and world cities, *The Geographical Review*, 88, pp. 528–547.

GOH, C. T. (1988) Our national ethic, *Speeches: A Bimonthly Selection of Ministerial Speeches*, 12, pp. 12–15.

GOTHAM, K. (2002) Marketing Mardi Gras: commodification, spectacle and the political economy of tourism in New Orleans, *Urban Studies*, 39, pp. 1735–1756.

GRIFFITHS, R. (1995) Cultural strategies and new modes of urban intervention, *Cities*, 12, pp. 253–265.

HALL, P. (2000) Creative cities and economic development, *Urban Studies*, 37, pp. 639–649.

HALL, T. and ROBERTSON, I. (2001) Public art and urban regeneration: advocacy, claims and critical debates, *Landscape Research*, 26, pp. 5–26.

HANNERZ, U. (1993) The cultural role of world cities, in: A. P. COHEN and K. FUKUI (Eds) *Humanising the City? Social Contexts of Urban Life at the Turn of the Millennium*, pp. 67–84. Edinburgh: Edinburgh University Press.

HARVEY, D. (1989) *The Condition of Postmodernity*. Oxford: Backwell.

HARVEY, D. (1993) From space to place and back again: reflections on the condition of postmodernity, in: J. BIRD, B. CURTIS, T. PUTNAM *ET AL.* (Eds) *Mapping the Futures: Local Cultures, Global Change*, pp. 3–29. London: Routledge.

HILL, R. and FUJITA, K. (2003) The nested city: introduction, *Urban Studies*, 40, pp. 207–217.

HO, K. C. (2002) Globalization and Southeast Asian urban futures, *Asian Journal of Social Science*, 30, pp. 1–7.

HUBBARD, P. (1996) Urban design and city regeneration: social representations of entrepreneurial landscapes, *Urban Studies*, 33, pp. 1441–1461.

JACOBS, J. M. (1996) *Edge of Empire: Postcolonialism and the City*. London: Routledge.

JANSSON, A. (2003) The negotiated city image: symbolic reproduction and change through urban consumption, *Urban Studies*, 40, pp. 463–479.

KELLY, P. F. (1997) Globalization, power and the politics of scale in the Philippines, *Geoforum*, 28, pp. 151–171.

KONG, L. (2000) Introduction: culture, economy, policy: trends and developments, *Geoforum*, 31, pp. 385–390.

KONG, L. and YEOH, B. S. A. (2003) *The Politics of Landscape in Singapore: Constructions of "Nation"*. Syracuse, NY: Syracuse University Press.

KUSNO, A. (1998) Beyond the postcolonial: architecture and political cultures in Indonesia, *Public Culture*, 10, pp. 549–575.

KUSNO, A. (2004) Whither nationalist urbanism? Public life in Governor Sutiyoso's Jakarta, *Urban Studies*, 12, pp. 2377–2394.

LEE, Y. S. (2001) Speech by Minister for Information and the Arts, on the completion of the Renaissance City Report, delivered in Parliament on 9 March 2000, Press Release (http://www.mica.gov.sg/pressroom/press_000309.htm).

LEE, Y. S. and YEOH, B. S. A. (2004) Introduction: globalisation and the politics of forgetting, *Urban Studies*, 41, pp. 2295–2301.

LEES, L. (2001) Towards a critical geography of architecture: the case of an ersatz colosseum, *Ecumene*, 8, pp. 51–86.

LEY, D. (1996) *The New Middle Class and the Remaking of the Central City*. Oxford: Oxford University Press.

LEY, D. and OLDS, K. (1998) Landscape as spectacle: world's fairs and the culture of heroic consumption, *Environment and Planning D*, 6, pp. 191–212.

MCCARTHY, J. (1998) Reconstruction, regeneration and re-imaging: the case of Rotterdam, *Cities*, 15, pp. 337–344.

MORRISON, Z. (2003) Recognising 'recognition': social justice and the place of the cultural in social exclusion policy and practice, *Environment and Planning A*, 35, pp. 1629–1649.

OLDS, K. (1995) Globalization and the production of new urban spaces: Pacific Rim megaprojects in the late 20th century, *Environment and Planning A*, 27, pp. 1713–1743.

OLDS, K. (1997) Globalizing Shanghai: the "global intelligence corps" and the building of Pudong, *Cities*, 14, pp. 109–123.

OLDS, K. (1998) Globalization and urban change: tales from Vancouver via Hong Kong, *Urban Geography*, 19, pp. 360–385.

OLDS, K. and YEUNG, H. (2004) Pathways to global city formation: a view from the developmental city-state of Singapore, *Review of International Political Economy*, 11, pp. 489–521.

PARENTEAU, R., CHARBONNEU, F., TAAN, P. K. ET AL. (1995) Impact of restoration in Hanoi's French colonial quarter, *Cities*, 12, pp. 163–173.

PAUL, D. (2004) World cities as hegemonic projects: the politics of global imagineering in Montreal, *Political Geography*, 23, pp. 571–596.

POW, C. P. (2001) *Urban entrepreneurialism and downtown transformation in Marina Centre, Singapore: a case study of Suntec City*. MA thesis, National University of Singapore.

POW, C. P. (2002) Urban entrepreneurialism and downtown transformation in Marina Centre, Singapore: a case study of Suntec City, in: T. BUNNELL, L. DRUMMOND and K. C. HO (Eds) *Critical Reflections on Cities in Southeast Asia*, pp. 153–184. Singapore: Times Academic Press.

SASSEN, S. (2000) Analytic borderlands: economy and culture in the global city, in: G. BRIDGE and S. WATSON (Eds) *A Companion to the City*, pp. 168–180. Oxford: Blackwell.

SEO, J. K. (2002) Re-urbanisation in regenerated areas of Manchester and Glasgow: New residents and the problems of sustainability, *Cities*, 19, pp. 113–121.

SHATKIN, G. (1998) 'Fourth world' cities in the global economy: the case of Phnom Penh, Cambodia, 22, pp. 378–393.

SILK, M. (2002) 'Bangsa Malaysia': a global sport, the city and the mediated refurbishment of local identities, *Media, Culture and Society*, 24, pp. 775–794.

SRIVASTAVA, S. (1996) Modernity and post-coloniality: the metropolis as metaphor, *Economic and Political Weekly*, 17 February, pp. 403–412.

TEEDON, P. (2001) Designing a place called Bankside: on defining an unknown space in London, *European Planning Studies*, 9, pp. 459–481.

TEO, P. (2003) The limits of imagineering: a case study of Penang, *International Journal of Urban and Regional Research*, 27, pp. 545–563.

VALE, L. J. (1992) *Architecture, Power and National Identity*. New Haven, CT: Yale University Press.

WAITT, G. (1999) Playing games with Sydney: marketing Sydney for the 2000 Olympics, *Urban Studies*, 36, pp. 1055–1077.

WATERMAN, S. (1998) Carnivals for elites? The cultural politics of arts festivals, *Progress in Human Geography*, 22, pp. 54–74.

WATKINS, H. and HERBERT, D. (2003) Cultural policy and place promotion: Swansea and Dylan Thomas, *Geoforum*, 34, pp. 249–266.

YANG, M. C. and HSING, W. C. (2001) Kinmen: governing the culture industry city in the changing global context, *Cities*, 18, pp. 77–85.

YEOH, B. S. A. (1999) Global/globalizing cities, *Progress in Human Geography*, 23, pp. 607–616.

YEOH, B. S. A. (2000) From colonial neglect to post-independence heritage: the housing landscape in the central area of Singapore, *City and Society*, 12, pp. 103–124.

YEOH, B. S. A. (2001) Postcolonial cities, *Progress in Human Geography*, 25, pp. 456–468

YEOH, B. S. A. (2004) Cosmopolitanism and its exclusions in Singapore, *Urban Studies*, 41, pp. 2431–2445.

YEOH, B. S. A. and HUANG, S. (1996) The conservation-redevelopment dilemma in Singapore: the case of the Kampong Glam Historic District, *Cities: The International Quarterly on Urban Policy*, 13, pp. 411–442.

YEOH, B. S. A. and KONG, L. (1994) Reading landscape meanings: state constructions and lived experiences in Singapore's Chinatown, *Habitat International*, 18, pp. 17–35.

ZUKIN, S. (1991) *Landscapes of Power: From Detroit to Disney World*. Berkeley, CA: University of California Press.

Measure for Measure: Evaluating the Evidence of Culture's Contribution to Regeneration

Graeme Evans

Introduction

My own blunt evaluation of regeneration programmes that don't have a culture component is they won't work. Communities have to be energised, they have to be given some hope, they have to have the creative spirit released (Hughes, 1998, p. 2).

As the above quote suggests, communities need hope—some would say 'trust' (Sennett, 1986)—in the process and outcomes of regeneration, not least since this instrumental process is controlled largely from the 'outside'. This sentiment can perhaps be read in two ways: culture is a critical aspect of mediating and articulating community need, as development is planned and takes shape, through culture's potential to empower and animate. This should in turn lead to participation in, and ownership of, regeneration by the residents and other beneficiaries in an area. Alternatively, culture-led regeneration can be used as a 'sop' to distract attention from the underlying power over place that finally manifests itself in the type of projects and landscapes created and imposed on communities and sites undergoing regeneration. As Klunzman (2004, p. 2) succinctly put it: "Each story of regeneration begins with poetry and ends with real estate".

The extent to which cultural facilities and programmes positively contribute to the regeneration of areas and neighbourhoods which have been subject to economic and physical decline, and multiple social problems—unemployment, poverty, crime, poor amenities, education and housing—has become a more central concern of governments and regeneration intermediaries. This is particularly so in view of the duration of this phenomenon, its replication in post-industrial and developing cities world-wide and the growing call for evidence to support the claims which are made by city and cultural organisations in their pursuit of substantial capital funding and leverage.

The opportunity provided for a longer view of culture and regeneration projects and strategies and the evaluation of their success in both cultural and regeneration terms, is the subject of this paper. This draws on a review of evidence—how that evidence has been derived and the evaluation and measurement of impacts undertaken in both academic and policy spheres. This review is based in part on a study initiated by the UK Department for Culture Media and Sport, which, like its predecessors and equivalents in other countries, has promoted an urban renaissance through the arts and creative industries.[1] This panacea is viewed as one of the few remaining strategies for urban revitalisation which can resist (or embrace) the effects of globalisation and capture the twin goals of competitive advantage and quality of life which culture, somewhat optimistically, might offer. The current cultural resurgence has also been fed by Porter's 'new economic model' of city competitiveness (1995) and 'lifestyle' indices of diversity, the creative milieu and 'class' (Landry, 2000; Hall, 1998; Florida, 2004), as essential ingredients in city survival and growth. This "rhetoric has entered the vocabularies of local cultural policymakers and city boosters alike" (Stevenson, 2004, p. 119).

The following discussion introduces the growing demands for evidence-based policy evaluation in this field, including the types of reporting which arise in the regeneration process and promotion of major cultural facilities. A typology of the main approaches to culture and regeneration is then summarised, with 'exemplar' arts and urban regeneration schemes cited in order to illustrate particular impacts, and the 'counterfactual' where evidence is contradictory to the official discourse. The next section reviews a range of specific impacts and measurements used to evaluate the effects of culture and regeneration in physical, economic and social terms, with examples of some of the evidence arising and its limitations. The paper concludes with an assessment of the gaps in evidence, with suggestions for a more integrated approach to measuring multiple regeneration outcomes. An underlying argument throughout is that the attention to the high-cost and high-profile culture-led regeneration projects is in inverse proportion to the strength and quality of evidence of their regenerative effects. This is in part due to problems of measurement and evaluation criteria not being established either *a priori* or consensually, but also to the fact that, like hallmark event projects, major culture-led regeneration schemes are not wholly grounded or rational-decision-based. They rely more on (blind) faith, "pork barrel politics" (Sudjic, 1993, p. 31) and constructed visions which appear not to look beyond the short-term physical impacts and landscapes they create.

Search for Evidence

Writing over 10 years ago, Lim concluded that there was "a need to sort out the hype from the substance" in the claims commonly made for culture-led urban revitalisation, with the more positive results achieved by culture-based regeneration

> tending to be too general in content and conclusions, a situation clouded by the fact that there are no clear guidelines as to how the effects of these developments should be evaluated (Lim, 1993, p. 594).

Bassett, writing in the same year, cautioned that it is important not to exaggerate the economic impacts of these (culture-led) strategies; that smaller cities could not emulate the success of major cultural centres; and that, fundamentally,

> Cultural regeneration is more concerned with themes such as community self-development and self-expression. Economic regeneration is more concerned with growth and property development and finds expression in prestige projects and place-marketing. The latter does not necessarily contribute to the former (Bassett, 1993, p. 1785).

Reconciling the social with the economic and physical outcomes of regeneration has therefore been a challenge which more recent culture-led regeneration projects are expected to meet and by which established schemes should now be judged.

At a conference entitled 'Building Tomorrow: Culture in Regeneration' held in Salford, north-west England, in February 2003, participants from national and city government and cultural agencies made a call for greater evidence to support the claims commonly made for the ways in which arts and cultural activity contribute to successful regeneration (DCMS, 2003). In particular, a concern was expressed for a 'joined-up approach'—a term coined to reflect the fragmented nature of government policy and departmental working, and between tiers of government, such as local, regional, national—and a longer-term perspective on the social and economic impacts that emerge over time. Greater emphasis on measurement and the quality of evidence itself was sought, recognising that

> The distinct lack of, and commitment to, in depth research into this issue creates a situation in which policymakers are unable to draw an evidence base upon which to make key decisions in the application of culture-led regeneration strategies (Bailey *et al.*, 2004, p. 47).

In many respects, the call for 'hard evidence' and measurement tools is part of the larger question of how regeneration itself is measured, how long should it 'take' and what makes for successful intervention in meeting policy objectives and community need—or, more fundamentally, how choices over development are made and evaluated and 'Pareto effects' are distributed and felt at a local level. A comprehensive review of urban policy, public choice and development appraisal is beyond the scope of a single article—although all require consideration in a critique of culture and regeneration. As Hall observes

> Moments of civic transformation tend to get portrayed in overly simplistic terms as seamless and unproblematic. The reality is much more messy (Hall, 2004, p. 63).

Evidence-based evaluation of urban policy and practice therefore needs to address a number of 'wicked problems' (Harrison, 2000), to which culture adds yet a further dimension.

Evidence-based Policy

Governments now refer to the need for 'evidence-based' policy-making and evaluation (PMSU, 2004), which can be interpreted on the one hand as rejection of, or at least disquiet with, simplistic ideological principles and more grand theories and, on the other, as a recognition that public policy interventions require robust testing and greater assessment of their 'fitness for purpose' and operational effectiveness in meeting policy objectives. This is seen as a necessity as competing needs and aspirations, opportunity costs and a more heterogeneous populace (Worpole and Greenhalgh, 1999, p. 38) demand more transparent 'evidence' of what works and where public intervention is good 'value for money', or not. The political imperative for evidence is therefore all-pervading, generating guidance and systems for the measurement of performance and impacts, and a range of quantitative indicators against which, in this case, regeneration programmes can be compared (ODPM, 2001, 2003). However, a recent review of the evaluation of social regeneration programmes called for a shift of focus *away* from evidence-based policy and practice, to building knowledge over time, drawing together local experience, research findings and, critically, a better understanding of trade-offs and political imperatives (Coote *et al.*, 2004). These latter aspects have been little considered and understood in the evaluation of culture-led regeneration to date and, therefore, the nature of 'evidence'—its perspicacity, and the need for a more grounded theory—emerges from this critical review.

Types of Reporting

Evidence in one sense can be compared with 'scientific proof', such as in peer-reviewed published research. However, in the fields of regeneration and cultural development, the outcomes and value systems/judgements have been neither explicit nor established *a priori*. One example of this is the limited number of government or sponsor assessments which present robust evidence on regeneration effects generally—a long-term and multifaceted process itself—and even less where cultural impacts within regeneration are concerned (DCMS, 2000; 2004; ODPM, 2003). In a review of 'evidence', the source of such material and its efficacy also present problems, due to the nature—political, entrepreneurial and contested—of major regeneration projects typically taking place over several years, durations which exceed political terms of office. This also fuels a lack of transparency within the development and decision-making process, as risk and criticism are minimised and projects are continually 'negotiated'—one reason for common budget overruns (Evans, 2001a, 2003). As Bassett (1993, p. 1785) observed, "the problem in the process of alliance-building [is that] a critical or oppositional aspect of cultural development is lost". Published evidence of culture's contribution to regeneration is usually presented in one of the following types of report. Some are evidence-based, most however, are not—starting with the most common.

1. Advocacy and promotion. Reports and prospectuses are often produced during the feasibility, development and initial impact phase, or to justify further resources and support. Typically, such material is presented in the form of promotional, PR and descriptive case studies for media and public consumption, and design masterplans. They are also used to report on and 'celebrate' major programmes, as they move to the next phase— for example, *Birmingham's Renaissance* (BCC, 2003), Barcelona's *Universal Cultural Forum* (2000), *Building Culture Downtown* (National Building Museum, 1998).

2. Project assessment. These are normally produced for internal (management) and external (state/funder) use. This type of report typically concentrates on financial and user-related outputs, such as income and expenditure, audience/visitor numbers, direct employment—i.e. resident organisations (such as an orchestra) and construction activity. They tend not to evaluate the process or outcomes of the project, or profile beneficiaries (or 'non-users'), or the user experience. They are used principally by the organisation and its funders in annual assessment and are rarely published, although often publicly funded. Project assessment is also normally carried out once, post-completion, whether a capital project (building) or activity programme (education programme, event/festival). Regenerative effects (if any) are therefore subsumed into the facility or organisation's overall performance.

3. Project evaluation. The focus in this case is as much on the process employed to plan and deliver a project as on the 'results'. They may include quantitative and/or qualitative evidence. The most common forms of data collection are questionnaires, unstructured interview/focus groups and participant observation. The evaluation may be of one project only or of a group of projects (see below) whether locally (regeneration area) or nationally (arts education programme). The evaluation may be carried out by the organisation itself or with the support of an external evaluator. It is recommended that evaluation is integrated from the outset of a project (baseline), undertaken during, on completion and post-completion (Jackson, 2005). Evaluation methodology has developed substantially in the past decade (Evans and Shaw, 2001a, 2001b) in the cultural and regeneration spheres, drawing particularly from environmental health (Bowling, 1997), crime prevention, urban design and quality of life measurement (Rapley, 2001), with an interdisciplinary focus on process and participation.

4. Programme evaluation. Wider programme evaluation is undertaken of schemes made up of separate projects with common aims, or typically part of a single initiative or funding programme. These can draw on project evaluation techniques (see above), but programme evaluation is also likely to entail standard output criteria (including PIs; see below)—for example, in regeneration programme assessment and grand-aid schemes. In the latter case, a comparative framework is used to assess individual projects as part of a wider initiative, whether local, national (Lottery), or transnational such as EU-funded programmes—for example, *European Capitals/Cities of Culture* (Palmer, 2004), *Culture and Neighbourhoods* (Bianchini and Ghilardi, 1997) and regional development funding of culture (Evans and Foord, 2000). Regeneration design and building types have been the subject of various published collections, notably waterfronts (Marshall, 2001; Wang, 2002), the reuse of heritage buildings (Bordage, 2002) and cultural facilities, particularly museums and galleries.

5. Performance Indicators (PIs). PIs are used to compare actual performance against targets and comparative standards, which are quantitative, benchmark and service-provision based. They therefore measure inputs (resources), throughput (capacity, attendance) and final outputs (productions, population penetration, such as frequency), but not process or outcomes: "the functional principle of organisation has bequeathed us any number of measures of activity but very few measures of outcome" (Perri 6, 1997, p. 65). PIs are applied more commonly in cultural organisations that are directly answerable to/ funded by government, such as national museums and galleries, and larger organisations funded on a regular basis by cultural agencies, such as national theatres and libraries (Arts Council, 1999; Evans, 2000). Per capita funding estimates are also used to make national comparatives, in pursuit of international benchmarks (DCMS, 2004, p. 75; Feist *et al.,* 1998). The growing use of quality of life indicators (see below) seeks to measure a range of environmental and liveability factors at local and national levels, including access to cultural amenities (DETR, 1998).

6. Impact assessment. These studies look at the likely or actual impact of an activity on a particular location/site, community or economy—typically economic impact, environmental impact (EIA), health impact, transport and tourism impacts—but these are seldom combined. Impacts are quantified wherever possible (or ignored where not) and intangible effects are translated numerically through the use of proxy measurement, such as cost–benefit analysis (CBA), contingent valuation and willingness to pay for otherwise 'free' activities (such as parks, museums and libraries). However, as Matarasso warns

> In a world of numbers and quantification, if there are no indicators to assess the value of activities, feelings or relationships, these things—however real—have no legitimacy (Matarasso, 1996, p. 1).

Secondary multipliers are also commonly used in economic and environmental impact studies. The use of formulaic impact methods, including disaggregated visitor and economic data, are seldom representative and are often out of date and not sectorally derived (Evans, 1998). Impact studies are also undertaken for large or environmentally sensitive schemes under national planning, European/EU and World Bank regulations, or are commissioned by local authorities, developers and investors. Full-blown cost–benefit and economic impact studies are seldom applied to culture-led and mixed (public–private) regeneration schemes and, where they are, there is little evidence of their *post hoc* evaluation or longer term reassessment. As a UK Treasury review of capital projects noted, the value of long-term benefit needs to be brought into the appraisal process (Treasury, 2003). Valuing benefits also requires greater consideration since, as the review also concluded, this aspect has been done poorly in the past, with many projects not describing and managing the realisation of benefits and overoptimism in projections, particularly capital costs.

7. Longitudinal impact assessments. Unlike one-off impact assessments, these take a baseline position and compare impacts over time or at least two points in time, in some cases mapping attitude and perception changes of residents, users/visitors, as well as more quantitative effects such as visitor levels, demographic change and economic/employment impacts. This model is used, like evaluation, both for individual projects and for programmes of activity. These are rare, often involving research centres and national/European comparative studies.

Longitudinal Studies

Regional city examples include longitudinal studies at Glasgow University (CCPR; www.gla.ac.uk), where a retrospective assessment of Glasgow's 'City of Culture' and successive event-based cultural regeneration is in progress, and Northumbria University (CISIR, see Bailey *et al.*, 2004) with a 10-year impact assessment of the quayside regeneration on Tyneside, or 'NewcastleGateshead' as these two divided cities have now been rebranded in order to bridge their historic divide and create a sense of place. In London and Toronto, a 3-year international comparative *Creative Spaces* study (LDA, 2004) is also underway, with numerous European networks of cultural development (such as ATLAS, CIRCLE, Budapest Observatory) undertaking cross-national studies on arts development and cultural funding. The impact of new transport insfrastructure on cultural activity and regeneration has also recognised the importance of access to positive regenerative effects. For example, the extension of the Jubilee Line underground line in London involved a longitudinal impact study of the effects of new stations on visitors to new and established cultural facilities (Evans and Shaw, 2001b).

The compilation of various indices of creativity and city growth is associated in North America with Florida (2004) and Nichols Clark (2004) and see Gertler *et al.* (2002) on Canada, although these do not directly address regeneration impacts. In some respects, these essentially lifestyle rankings which correlate selected location advantages with creative milieus, are also associated with socio-cultural and spatial inequalities and gentrification effects. They are also not necessarily linked to higher productivity or innovation, despite the knowledge-based creative industries they supposedly represent (DTI, 2004; Simmie, 2001).

Longitudinal impact studies have also been undertaken of public and related environmental art (Shaw, 1990; Hamilton *et al.*, 2001). In a study of public art and commercial property, Roberts and Marsh (1995, p. 192) found that "the image or attractiveness of a development was a significant factor in an occupier's choice of building", although rental cost, location and quality were more important. However, as Ward Thomson *et al.* (2004) point out, most texts on this subject have little or no consideration of evaluation, confirming a resistance by practitioners to this process, despite the claimed benefits of 'public' (*sic*) art (Selwood, 1995). Hall and Robertson (2001) suggest that much public art criticism, although avowedly about the reception of public art, is actually written from within a 'productionist framework'—i.e. by artists and arts administrators who fail to say very much about the public reception of the work.

Major annual festivals such as Edinburgh (Gratton and Taylor, 1995; Prentice and Andersen, 2003) and the Notting Hill Carnival (LDA, 2003) have also been the subject of longitudinal impact studies, including the European City of Culture (ECC) programme. Originating in Athens in 1985, this has supplemented the International EXPO series originating in the early 19th century, with host cities using these events as part of their international profile-raising and longer-term regeneration of run-down areas, notably Seville (EXPO 1992), Lisbon (City of Culture 1994, EXPO 1998), Rotterdam (City of Culture 2001), Barcelona (1992 Olympics, UNESCO Cultural Forum 2004) and Liverpool in the run-up to Capital of Culture in 2008. Cultural festivals now feature in major sporting events which combine area regeneration, such as the 2002 Commonwealth Games, Manchester and contemporary Olympic bids. Although the ECC programme was a self-conscious extension of the 'European project', the first City of Culture of the Americas was held in Merida, southern Mexico, in 2000. The recent assessment of ECCs (Palmer, 2004) found, however, that they have too often focused their efforts on funding one-off events and projects, with little time and investment given to the future.

Cultural Pessimism?

Where research on the arts and urban regeneration has featured in academic articles, these tend to be either descriptive and uncritical case studies, or highly critical (but lacking in robust empirical evidence), displaying a 'culture of pessimism'. This in contrast to both official and media discourses, and the promotional literature which surrounds these major schemes, with the exception of projects which run into financial crisis. The latter can occur: at the feasibility stage, with schemes that never see the light of day (for example, Cardiff Bay Opera House); when under construction as budgets are exceeded or designs fail (many 'grands projets' in France and Britain, Olympic stadia, such as in Athens 2004); after opening, when visitor targets and income are not met (for example, National Centre for Popular Music, Sheffield); or when management/leadership fail (for example, Prado, Madrid). Even 'successful' cultural facilities, judged so in the short term, suffer from leadership change, with artistic directors leaving within a year or so of opening. This phenomenon can be seen in 'provincial' cultural facilities importing their first directors and 'talent' from the capital city (for example, Paris; see Negrier, 1993) or from overseas, reflecting the international circuit which these cultural intermediaries now inhabit, along with the star international architects who design their showcase buildings (Evans, 2003).

Although not normally presented as 'evidence' in policy evaluation and impact assessment, in many respects academic discourses act as a counterbalance or response to the 'official' stories surrounding major regeneration and cultural development projects, including media depictions and campaigns. In some respects, the lack of 'hard evidence', of access to detailed data, decision-making processes and the basis to measure impacts over time, have limited more 'scientific' analysis, leaving little scope, outside political and policy discussions and micro-level case studies. These include more technical quantitative impact assessments (economic/econometric) and area and sectoral case studies of a particular cultural activity—for example, theatre quarter, fashion industry, heritage tourism and creative production 'clusters' (Evans, 2004a; Montgomery, 2003)—within a larger regeneration programme or site. As Bassett et al. point out, "detailed case studies are useful as windows onto local governance, helping to illuminate deeper aspects of local politics and power structures" (Bassett et al., 2002, p. 1773).

Recent special issues on the culture and regeneration phenomenon, including the related area of creative cities/industries, reflect the growing interest across discipline and subject areas—for example, journals of cultural policy (*IJCP*, 2004), urban and regional research (*IJURR*, 2003), *Local Economy* (2004) and *Cultural Studies* (2004). More literature exists in the related fields of leisure and tourism studies, where impact studies (as opposed to evaluation) are common, particularly in the growing area of cultural/heritage and urban tourism. Whilst in some cases academic publications may also provide alternative perspectives, their currency limits any real impact on the regeneration process and the 'promotions' and 'protests' which surround them. Methodologies which bring together approaches across anthropology, cultural and urban studies/sociology (Stevenson, 2004) and apply these to evaluation models which can measure social, economic and physical change, are yet to be developed, although in culture and regeneration this is what the phenomenon demands.

The more pessimistic discourses also reflect a rejection, or at least suspicion, of the effects of post-industrial (post-Fordist) urban development and the commodification of place (Harvey, 1993, p. 8; Robins, 1996; Kearns and Philo, 1993). A growing literature on city place-making and branding (Kavaratzis, 2004; Hannigan, 2003; Ward, 1998; Hauben *et al.*, 2002), which draws on marketing and product life-cycle concepts, in many ways cements this convergence of culture and commerce, and therefore of culture and regeneration, but without a deeper analysis of either concept or their relationship. As Paddison's earlier study of Glasgow's city reimaging campaign concluded, such marketing strategies tend to overlook the social and political implications they raise (Paddison, 1993, p. 339; and see Griffiths, 1993). An overconcentration on brand image also risks the inevitable process of 'brand decay' (Evans, 2003), which is seen in the reimaging and reinvestment in 'new' cultural facilities and experiences, in order to maintain visitor appeal and city marketing distinction. The demand for more robust evidence reflects in some quarters, resistance to the acceptance of the creative or 'thin air' economy (Leadbeater, 2000) as robust, or as an industry at all. Most criticism, however, is levelled at the gentrification effects which are associated with regeneration through what are seen as amenities and activities for the professional managerial classes (PMCs; see Ehrenreich and Ehrenreich, 1979; McGuigan, 1996) and who are therefore the disproportionate beneficiaries through participation, consumption and employment in this 'trade in signs and symbols' (Lash and Urry, 1994).

From Economic to Social Impacts and Quality of Life

Culture-led regeneration, or rather regeneration using cultural events and flagship projects, has also widened the rationale for cultural investment to include social impacts, in particular, arts-based projects which address social exclusion, the 'well-being' of city residents and greater participation in community life. As Betterton maintains

The focus on the economic benefits of the arts and urban regeneration was overstated in the 1980s ... The argument has now

shifted back towards more 'soft edged' rationales for cultural investment: cultural activity as one key indicator of a city's quality of life (Betterton, 2001, p. 11).

Physical regeneration has not been limited to building-based flagships and city-centre public realm schemes, but is increasingly seen through smaller public art projects and concern for design quality in the everyday environment

Regeneration is not simply about bricks and mortar. It's about the physical, social and economic well being of an area; it's about the quality of life in our neighbourhoods. In relation to the physical, this is as much about the quality of public realm as it is about the buildings themselves (ODPM, 2001, p. 3).

The arts have generated interest in regeneration through their symbolic potential, such as heritage and identity, assisting in change processes and cultural expression, and in reaching the parts which other regeneration activity does not reach. Examples include the use of heritage resources in developing greater social inclusion (Newman and McLean, 1998) and valuing identity amongst communities where historic industrial sites undergo culture-led regeneration (Bailey *et al.*, 2004). 'Public good' benefits arising, particularly from symbolic sites which are the subject of regeneration (Hayden, 1995), include their option, legacy and prestige values—where community members do not actually attend or directly benefit themselves (and even may have no interest in the culture on offer), but see value to others and also take pride in the development.

What are now looked for—and this distinguishes the position today from the 1980s—are the twin benefits of social cohesion and economic competitiveness and their interrelationship, through regeneration and related neighbourhood-based intervention (Boddy and Parkinson, 2004), seeking "Better engagement/consultation with local communities to improve ownership of the (cultural) project and (local) benefits"

(DCMS, 2003, p. 2). This confirms that, in measuring and evaluating regeneration programmes and culture-led regeneration, the tests of sustainability and distributive equity are now imperatives, suggesting that short-term impacts have not been sustained in the past and that social benefits have not been achieved, or have even been displaced by the gentrification associated with major redevelopment projects and high art venues. This is reflected in one view of the French *grands projets culturel*

> Whatever their value as architectural set-pieces, they are not the much-vaunted harbingers of a proclaimed urban renaissance. On the contrary, like circus games, they direct attention from the inexorable erosion of Paris and the brutal neglect of its suburbs (Scalbert, 1994, p. 20).

The contemporary adoption of the notion of 'social exclusion'—and associated Third Way policy responses (Stevenson, 2004)—originated in the housing estates of outer Paris and has informed cultural policy as much as broader social policy (EC, 1998) and urban regeneration objectives. Policy and project reviews which explicitly explore these social effects[2] include 'The Arts and Neighbourhood Renewal' (Shaw, 1999, 2003), the 'Evaluation and Social Impact of the Arts' (Matarasso, 1996, 1997); studies of arts and social inclusion in Australia, Canada and Scotland (Williams, 1996; Jeannotte and Stanley, 2002; Ruiz, 2004) and an assessment of the social impacts of lottery-funded projects in the UK (Evans and Shaw, 2001a; and see Jackson, 2000). The UK state-sponsored lottery celebrated its 10th anniversary in 2004 and has allocated over £15 billion to 'good causes' over this period, including major capital cultural, sport and millennium schemes—Britain's response to the '*grands projets*'. This has fuelled a roll-call of mega-projects which together have changed the landscape of cultural facilities in Britain at a scale not witnessed since the Victorian heyday and the inheritance from 19th- and early 20th-century international EXPOs and World Fairs. These buildings are too 'young' to be producing evidence of sustainable impact, although there is no shortage of claims for their *expected* impact. Much of the attention to evaluation has arisen from this massive programme, although this has been *post hoc* and limited to advocacy, project assessment and evaluation reporting, as outlined above. It is no coincidence that new European Union members look to state lotteries as a prime source of cultural investment (Bodo *et al.*, 2004), alongside the regional development aid which has benefited cultural projects in poorer industrial and rural regions of western Europe (Evans and Foord, 2000).

Models of Regeneration through Cultural Projects

The conflation of the social (*inclusion, liveability*) with the economic (*competitiveness, growth*), through physical redevelopment and architecture, reflects the current understanding of what site-based regeneration seeks to achieve. The term regeneration has been defined as the transformation of a place—residential, commercial or open space—that has displayed the symptoms of physical, social and/or economic decline

> breathing new life and vitality into an ailing community, industry and area [bringing] sustainable, long term improvements to local quality of life, including economic, social and environmental needs (LGA, 2000, p. 3).

Evidence of regenerative effects can therefore be sought where culture is a driver, a catalyst or at the very least a 'key player' in the process of regeneration or renewal. Three models through which cultural activity is incorporated (or incorporates itself) into the regeneration process can be distinguished over this period: culture-led regeneration, cultural regeneration and culture *and* regeneration, although these are not necessarily mutually exclusive, particularly over the longer term.

Culture-led Regeneration

In this model, cultural activity is seen as the catalyst and engine of regeneration—epithets of change and movement. The activity is likely to have a high-public profile and frequently to be cited as the sign or symbol of regeneration—most notably, the cultural flagship or complex. The activity might be the design and construction (or reuse) of a building or buildings for public or mixed use; the reclamation of open space (for example, garden festivals, EXPO sites); or the introduction of a programme of activity which is then used to rebrand a place, notably arts 'festivals', events and public art schemes. What these cultural interventions share is a claim for a uniqueness which 'non-cultural' regeneration such as the less glamorous housing, office, retail and site reclamation developments lack, a means for creating (or rediscovering) distinctiveness and for raising awareness and excitement in regeneration programmes as a whole.

Use and misuse. The phrase 'culture-led regeneration' is now commonly (mis)used where there is a high-profile arts facility in area regeneration, but in most cases this is a visible but less significant element in a wider and longer-term development scheme and investment programme, used to front, but not necessarily, drive, property and other economic development. Even in cases where a new cultural flagship dominates the external image and landscape, such as Guggenheim Bilbao, Spain, the Lowry, Salford, and Baltic/Sage Gateshead, UK, this belies major (and prior) investment in land preparation and transport infrastructure—air, road, rail/light rail, metro, as well as upmarket housing and hotels—and the upgrading of existing cultural facilities. Regenerative effects, in distributive and sustainable terms, on the other hand may be low particularly where economic leakage is high and regeneration activity and economies lack diversity. And as Giddens remarked

> Money and originality of design are not enough … You need many ingredients for

big, emblematic projects to work, and one of the keys is the active support of local communities (Giddens; quoted in Crawford, 2001, p. 2).

However, a feature of many flagship developments has been resistance by, or bypassing of, local communities (MacClancy, 1997; Plaza, 2000; Rodriguez *et al.*, 2001), with the legacy of event-based regeneration not delivering sustained benefits or ownership by residents, as in post-EXPO/Olympic Montreal (Kroller, 1996) and, more recently, Barcelona, Bilbao, Lisbon, Salford and Sydney (Garcia, 2004; KPMG, 1998).

Cultural Regeneration

In this model, cultural activity is more integrated into an area strategy alongside other activities in the environmental, social and economic sphere. Examples include the city of Birmingham where, at an early stage of the city's 'renaissance' (BCC, 2003), 'culture' was incorporated with mainstream policy, planning and resourcing through the council's joint Arts, Employment and Economic Development Committee, and in the 'exemplar' cultural city Barcelona which early on took an urban design, cultural planning and creative quarter approach, which is still recreating itself through the further expansion from the old city out to the former Olympic village site and declining Poblanou industrial district. This former manufacturing area on the city fringe is now targeted as a creative industries quarter, linking the old, overheating city to the expanding waterside commercial development promoted through a UNESCO 'EXPO' site. According to Gdaniec, "The redevelopment plan for Poblanou can be regarded as a model of how cultural production can flourish in a marginal area", but, as he goes on to admit,

> Urban regeneration combining culture can result in fragmented and unreal spaces, as well as contested space and culture … in Poblenou, speculation and quasi-exclusion of locals from the new housing (Gdaniec, 2001, p. 387).

Creative city exemplars? Birmingham, which looked to Barcelona and North American cities such as Chicago for inspiration for their prestige city-centre redevelopment plans, incorporating major arts and events facilities, public art and landscaping schemes, also presented an early example of the divided 'event city'. Criticisms by local and central government and in the press were made of the diversion of public spending from mainstream education and social programmes (Loftman, 1990). This pattern of promotional and celebratory reports issued by the agents and promoters of regeneration, followed by more dismissive critical responses in the academic literature, typifies this field. Moreover, the perspective of local communities is less apparent and visible, at least from the outside, but some communities in self-styled cultural cities tend to remain 'outside'— perceiving these city-centre and new cultural spaces as not for them and 'inaccessible'— for example, the residents of Barcelona's El Raval (Miles, 2004; Ulldemolins, 2000); Birmingham's afro-caribbean youth (Dudrah, 2002; Symon, 1999); Chicago's south-siders and the occupants of Salford's precinct housing estates, where the metro train which links the Lowry cultural quarter to Manchester, does not reach.

Exemplars feature highly in city cultural and regeneration plans and have been a prime reference for cities seeking to emulate their 'success'. This goes as far as replicating design schemes, public art installations, and themes, such as modelling Barcelona's museum of contemporary art (MACBA, El Raval) on Pompidou—the "Beauborg of Barcelona" (Balibrea, 2001, p. 198); and in London, Covent Garden's piazza and mixed-use development mirroring Quincy market in Boston, as well as port cities such as Madeira and Montreal looking to emulate Barcelona's waterfront redevelopment.

This approach is also closely allied to the 'creative city' model of urban cultural policy and regeneration (Landry and Bianchini, 1995) and one to which the current culture and design-led city visions have turned

(Landry, 2000), including those where cultural flagships have failed to sustain or fulfil their promise in social and/or in economic regeneration terms. This is seen in the renaissance plans of industrial and port cities such as Bradford, Barnsley, Salford and Liverpool in the UK; Valencia, Marseilles and Rotterdam on the continent and waterfronts in Toronto and Montreal. From a near-historical perspective, the reality that many of these cities are on their second or third cultural investment, place-making and economic strategy, is seldom reflected in their current promotional literature or in the critiques and assessments of their latest plans. The opportunity for evaluation informing future development— what works, what hasn't—appears to be ignored in this revisioning process—this despite many years of public regeneration and regional development subsidy and, in some European cases, successive funding programmes which have run continually for several years (Evans and Foord, 2000).

Culture and *Regeneration*

In this 'model by default', cultural activity is not fully integrated at the strategic development or master planning stage, often because the responsibilities for cultural provision and for regeneration sit within different departments or because there is no 'champion'. Such interventions are often small: a public art programme for office development, once the buildings have been designed; a heritage interpretation or local history museum tucked away in the corner of a reclaimed industrial site. In some cases, where no planned provision has been made, residents (individuals or businesses) and cultural organisations may respond to the vacuum and make their own interventions—commissioning artists to make signs or street furniture, recording the history of their area, setting up a regular music night and so on. Although introduced at a later stage, cultural interventions can make an impact on the regeneration process, enhancing the facilities and services that were initially planned. It is important to note that the lack of discernible cultural activity or provision within a regeneration scheme does not necessarily mean that cultural activity is absent, only that it is not being promoted (or recognised) as part of the process.

Retro-fitting culture. Reasons why culture is frequently an 'add-on' rather than an integral part of a scheme include the fact that the local authorities and partnership bodies responsible for regeneration schemes are rarely structured to facilitate collaboration between those responsible for regeneration and those responsible for cultural activity and they may not naturally think of themselves as collaborators. The other common reason is the lack of a champion with experience of what cultural activity can contribute to regenerative projects. Leadership appears to be a fundamental ingredient for credibility to be established at city, national and international levels. The absence of a powerful voice can therefore disadvantage the less well-heeled and less connected groups and communities.

Frequently, regeneration programmes are developed without reference to, or inclusion of, incumbent arts and cultural groups, or past heritage associations/communities. This arises due to the different nature and perspective of the 'regenerators' and community-based activity (including municipal and 'amateur' arts) and the preference for the 'new' (flagship, public art, employment, residents, visitors) over the 'old'. Indeed, in the areas which are the subject of extensive regeneration, it is presumed that quality of life and, by association, indigenous culture, is poor and needs 'improving'. The factors that lead to the creation of cultural flagships, mega-events and related arts programmes in practice can minimise public choice or the more objective analysis of *cultural* impacts from a regeneration perspective and vice versa. Evaluation of the processes (see below and Table 1) which measure decision-making and stakeholder consultation is therefore important, since this will influence *a posteriori* assessment of community involvement, ownership and the success of a particular scheme.

Cultural Planners

These may be extreme examples (although feted as 'exemplars') and the conditions to improve this situation are evident with more integrated and inclusive cultural planning through guidance and toolkits on local cultural strategies (DCMS, 1999; Evans, 2001a). In this sense, lessons are being learned, both tactically and strategically (Landry, 2000). But the power of capital over culture should not be understated, however, liberal or benign a particular regeneration regime may present itself. Community consultation is a prerequisite and tool which developers and their designers now employ, but evidence of the impact of such consultation in the final built schemes is less apparennt

> City elites have now learned how better to incorporate dissenting groups (middle class ones, at least) and manage potential conflicts more effectively (Bassett *et al.*, 2002, p. 1774).

The pluralist model of regime theories (Stoker, 1995) suggests that, through multiple stakeholders, power over decision-making and resources is more equitably distributed, that minority and small, special-interest groups can influence outcomes. Whilst the process may be consensual, the physical end-product may be less so—evidence of local and community (however defined) influence on the shape and content of cultural facilities within regeneration schemes is rare, whilst masterplanners, star architects and cultural intermediaries are brought in to create a vision of place. Writing on waterfront regeneration in Bristol, Bassett *et al.* (2002, p. 1774) acknowledge that "the final masterplan is still within the broad parameters laid down at the very beginning by the planning brief". McCarthy (2002) also concluded, in an assessment of Detroit's entertainment-led regeneration, that the governance context is all-important if this approach is to be at all successful and where culture and regeneration are used to encourage gentrification and urban resettlement, as Seo (2002) found, objectives of social cohesion and sustainability are compromised.

Table 1. An overview of the evidence of culture's contribution to regeneration

Physical regeneration	Economic regeneration	Social regeneration
Policy imperatives:		
Sustainable development	*Competitiveness and growth*	*Social inclusion*
Land use, brownfield sites	Un/Employment, Job quality	Social cohesion
Compact city	Inward investment	Neighbourhood Renewal
Design quality (CABE, 2002)	Regional development	Health and Well-being
	Wealth Creation	Identity
Quality of Life and Liveability	SMEs/micro-enterprises	Social Capital
Open space and amenity	Innovation and Knowledge	Governance
Diversity (eco-, landscape)	Skills and Training	Localism/Governance
Mixed-Use/Multi-Use	Clusters	Diversity
Heritage conservation	Trade Invisibles	Heritage ('Common')
Access and Mobility	(e.g. tourism)	Citizenship
Town Centre revitalisation	Evening Economy	
Tests and measurements:		
Quality of Life indicators	Income/spending in an area	Attendance/Participation
Design Quality Indicators	New and retained jobs	Crime rates/fear of crime
Reduced car-use	Employer (re)location	Health, referrals
Re-use of developed land	Public-private leverage/ROI	New community networks
Land/building occupation	Cost benefit analysis	Improved leisure options
Higher densities	Input-Output/Leakage	Lessened social isolation
Reduced vandalism	Additionality and	Reduced truancy and
Listed buildings	substitution	anti-social behaviour
Conservation areas	Willingness to pay for	Volunteering
Public transport/usage	cultural amenities/ contingent valuation	Population growth
	Multipliers—jobs, spending	
Examples of evidence of impacts:		
Reuse of redundant buildings—studios, museum/gallery, venues	Increased property values/rents (residential and business)	A positive change in residents' perceptions of their area
Increased public use of space—reduction in vandalism and an increased sense of safety	Corporate involvement in the local cultural sector (leading to support in cash and in kind)	Displacing crime and anti-social behaviour through cultural activity (for example, youth)
Cultural facilities and workspace in mixed-use developments	Higher resident and visitor spend arising from cultural activity (arts and cultural tourism)	A clearer expression of individual and shared ideas and needs
High density (live/work), reduce environmental impacts, such as transport/ traffic, pollution, health problems	Job creation (direct, indirect, induced); enterprise (new firms/start-ups, turnover/ value added)	Increase in volunteering and increased organisational capacity at a local level
The employment of artists on design and construction teams (Percent for Art)	Employer location/retention; Retention of graduates in the area (including artists/ creatives)	A change in the image or reputation of a place or group of people
Environmental improvements through public art and architecture	A more diverse workforce (skills, social, gender and ethnic profile)	Stronger public–private–voluntary-sector partnerships

(Table continued)

Table 1. *Continued*

Physical regeneration	Economic regeneration	Social regeneration
The incorporation of cultural considerations into local development plans (LPAC, 1990)	Creative clusters and quarters; Production chain, local economy and procurement; joint R&D	Increased appreciation of the value and opportunities to take part in arts projects
Accessibility (disability), public transport usage and safety	Public–private–voluntary-sector partnerships ('mixed economy')	Higher educational attainment (in arts and 'non-arts' subjects)
Heritage identity, stewardship, local distinctiveness/vernacular	Investment (public–private sector leverage)	Greater individual confidence and aspiration

Furthermore, the internationalisation of the masterplanning and regeneration process is no longer limited to start architects/construction firms and a mobile cultural élite, but also to cultural planners who operate globally, with a small number of ubiquitous arts consultants featuring in creative strategy initiatives world-wide (for example, the World Bank; see Evans, 2001b). Writing from one recipient region, Stevenson remarks: "The relevance to Australian cities and cultures of these European-inspired prescriptions is an open question" (Stevenson, 2004, p. 129). There is now evidence that cultural strategies developed by external 'catalysts' have actually reinforced spatial divides and social exclusion, particularly amongst cultural minority and social groups (Evans and Foord, 2003), or the aspirations of the creative district and economy raised unrealistically, setting up local agencies to fail after the roadshow has moved on. References to ethnic and regional culture in design, content and operation of these cultural facilities have a tendency to fall at the outset (for example, the Guggenheim, Bilbao, see Baniotopoulou, 2000, and Rodriguez *et al.*, 2001; the MuseumQuartier, Vienna, see Mokre, 1998), or all but disappear between the design concept stage, where community and political support (and minimisation of resistance) are required, and the final product. As Chang (2000) documented in the case of the 'Esplanade Theatres by the Bay' in Singapore, this $250 million 1800-seat concert hall and 2000-seat lyric theatre was supposed to house smaller studios and performance spaces for local groups, but these plans were eliminated early on. Arts practitioners expressed concern that with its mega-structures and high rentals, it will be amenable mainly to blockbuster events such as foreign pop concerts and Broadway shows, and less accommodating towards local and non-profit productions. A similar scenario is played out in mega-cities such as Shanghai, where Western-style regeneration is "sapping the city's own creativity" (Gilmore, 2004, p. 442; Wu, 2000).

Good and Bad Practice

All three of these culture/regeneration 'models' therefore provide examples of positive and negative effects. Their scope and motivation will also dictate the evaluation criteria and success factors which can be applied and the outcomes that might be expected from the cultural aspect of regeneration in each case. There are culture-led regeneration projects that have been too ambitious in their projections and landmark buildings that have failed to reach their targets (in terms of audience numbers, profiles and income generated) or secure community ownership. There are culture *and* regeneration projects in which arts programmes have been 'retro-fitted' to poorly conceived developments in an attempt to improve their appearance, to animate a place or to secure community involvement. The regenerative effect of *cultural* impacts also arises—the impact of cultural activity on the culture of a community, its codes of conduct, its identity—and notions of citizenship, participation and diversity. This approach to reshaping cultural landscapes is also consonant with Bennett's argument that culture itself should be thought of as 'inherently governmental', so that "culture is used to refer to a set of practices for social management deployed to constitute autonomous populations as self-governing" (Bennett, 1998, p. 884). Cultural governance in this sense is another factor which can be assessed and which might offer useful approaches to community engagement in the fraught regeneration process. How these models of regeneration through/with culture can all be identified, their effects measured and schemes evaluated, requires questions to be asked at various stages in the process and change assessed and attributed as impacts are felt.

Impact Measurement and Indicators of Change

The generic term 'impact study' is now widely used in relation to the 'contribution' or 'role' or 'importance' of cultural activity to another objective—in this case, to regeneration. Much of the literature on the contribution of culture to society now uses the language of impacts. Studies that look beyond the project itself traditionally use one (and seldom more than one) of social, economic and environmental impact, which is generally tested using particular measurements. Table 1 summarises the current imperatives that drive a range of public policy agendas—economic and social—and therefore area regeneration programmes, including those with a cultural element. The tests by which policies are measured in practice in terms of physical, economic and social change, are largely quantitative, including the familiar economic and environmental impact indicators, but also more qualitative evaluation, particularly in terms of behavioural effects, social capital and perceptions such as community safety and the socially constructed notions of exclusion, diversity and heritage (Andra, 1987). Environmental

or more broadly, 'quality of life' indicators (above) have developed from international policy initiatives such as Agenda 21 and its local application—LA21—and are now used in countries such as the UK and Canada as national and local benchmarks of liveability (DETR, 1998; FCM, 2001). These include social as well as environmental qualities, such as access to services, fear of crime and community cohesion. 'Culture' *per se* does not have an equivalent 'quality of life' indicator set, outside the aspirational declarations of 'cultural rights' (CLRAE, 1992) and 'common heritage' (Maastricht; see HMSO, 1993), although measures of social impacts through arts participation have been developed (Matarasso, 1996). Not surprisingly, the indicators most commonly referred to in linking the arts and regeneration are those now widely used in the context of neighbourhood renewal and social inclusion, quantifiable—essentially—by reduced levels of crime, increased health and well-being, increased educational attainment, reduced unemployment, greater community cohesion and improved environmental quality (DCMS, 2003, p. 2).

Some examples of the kind of evidence which arise and flow from these policy imperatives and measurements are then indicated in Table 1. These are not exhaustive nor ranked in any sense, since they will vary according to the nature and scale of regeneration undertaken, local conditions, 'history' and the objectives being pursued. As recent guidance on regeneration evaluation recommends

A pick and mix approach is required as there is no universally applicable set of indicators that will be appropriate for a particular intervention [and, in heritage and cultural impacts in particular] valuing in this area tends to be highly context specific (ODPM, 2003, p. 164).

This also suggests that standardised performance indicators and quantitative benchmarks are neither desirable nor useful measures in this situation.

Additionality

The nature and variability of the cultural element in regeneration projects, surprisingly receive little attention, outside the external design which dominates media and design-sector coverage. The cultural programme, its purpose, sustainability, mix and relationship with regeneration objectives, is treated independently—being largely the preserve of arts and cultural organisations and funders (including state cultural institutions and private patrons)—and therefore has a benign place in the overall regeneration scheme. The evaluation and risk assessment of cultural projects also varies widely between those projects based on existing and relocated arts organisations and collections (such as galleries and museums) or where there is a clear and authentic heritage or symbolic association, and those where the cultural facility or concept is new to the area. Whether a traditional cultural venue or an experiential visitor 'theseum', relying on latent demand is risky. However, in both cases, safe and incremental culture-led regeneration projects can fail to attract or maintain attention, whilst new cultural experiences in new locations can attract and maintain visitors, such as the Eden project in Cornwall, the BALTIC, Gateshead in England and Parc de la Villette on the outskirts of Paris.

The geographical focus of regeneration projects means that it is particularly important to assess displacement effects at the local, regional and national levels, particularly if the programme is substantial, or for major flagship projects. This is extremely hard to measure in practice through tests of additionality and substitution ('zero-sum game'; see Bianchini, 1991), so much so that such local and regional impact assessments are not feasible (Connolly, 1997), leaving this question to anecdotal and largely non-attributable effects of culture-led regeneration in one area, on another. The feasibility and evaluation studies undertaken pre- and post-regeneration projects ideally would attempt to measure these wider effects and the assessment of demand/need, which in the past has been based upon basic urban settlement planning norms of amenity and infrastructure levels required to support a largely homogeneous population. Parameters are influenced by accessibility and transport; however, where cultural activities, heterogeneity and more intangible experiences are concerned, quantitative planning models, as used for parks and sports facilities, are less useful and may actually reinforce exclusion (Evans, 2001a).

Public Realm and Urban Design

A particular paradox which has emerged from the cultural flagship phenomenon—apart from the search for distinctiveness through commissioning copycat designs/designers (Evans, 2003)—is in the evaluation of buildings and public realm schemes themselves. These images dominate the reporting and promotion of regeneration schemes and are critiqued in the architectural press and monographs on both buildings and city transformations. A typical image in these publications is the blue sky backdrop to a person-free building, providing an optimum view of the finished product. Such appraisals tend to make no reference to the regeneration context or outcomes from the building, leaving its aesthetic impact as the prime contributor. This is in contrast to what is now a succession of new flagship cultural buildings which have failed to meet their operational and user requirements, where form has undermined function, even in award-winning schemes (Evans, 2003). An observation from culture-led schemes world-wide is that, despite the architectural and media attention to the design experiments which house otherwise traditional performing and visual arts spaces and programmes, public preference is still strongly directed at the prosaic bridges, ferris wheels and waterfront boardwalks and the reuse of industrial structures (Bordage, 2002) from Tate Modern, London; the BALTIC, Gateshead, and their respective millennium bridges, to the 'MuseumQuartier', Vienna (Bogner, 2001).

Poorly designed and uninspiring 'municipal' interiors, expensive materials and fittings which require high maintenance, or fail (Guggenheim Bilbao), are less popular legacies which can limit regeneration effects over time. Likewise, public realm and spaces created in and around these projects, can be left over or unfinished. Given the scale of investment and expectations they demand, developing measurements of design quality requires greater consideration, not just of the designated public art schemes. Urban design quality indicators (DQIs) and techniques developed in landscape and pedestrian planning (Gehl, 2001) now draw on more social, obervational and qualitative assessment of the user experience. As Lefebvre observed, "The user's space is *lived*—not represented" (Lefebvre, 1974, p. 362). In the case of DQIs, this encompasses a wide range of 'users'—internal and external. Benefits which design indicators thus seek to measure include: identity/civic pride; place vitality; inclusiveness; connectivity; safety; and facilities and amenities, alongside the aesthetic design values and operational effectiveness over time (CABE, 2002; Carmona *et al.*, 2002).

Sustainability, in both design quality and integration terms, is therefore a fundamental success factor, which requires greater attention in evaluation and in learning from experiences elsewhere. Examples where repeated retro-fitting and makeovers take place are one indication of less sustainable regeneration schemes. Integration, which at the micro level includes the degree of mixed-use within a regeneration site and quarter, can be assessed in relation to existing urban areas, whether city centre or fringe. A particular feature of less successful waterfront, mixed-use and downtown regeneration schemes has been the failure to link with incumbent business and residential communities, or the creation of interstitial spaces which become unsafe or redundant—for example, waterfront and city centre/CBD-based regeneration schemes that remain empty and soulless much of the time, outside designated event usage. Mono-use complexes and/or those dominated at particular times by one user-group can also create exclusion and unsafe perceptions by other groups. This is becoming evident where strategies promote late-night opening and a club and drinking culture dominated by young people in city and town centres (Thomas and Bromley, 2000). Integration which translates into positive regeneration indicators therefore includes safety/natural surveillance; labour market access; disability access, diversity (ethnic, lifestyle, age-groups), as well as the production and consumption flows which accrue to more successful compact regeneration areas.

Conclusions

The evidence of regeneration using major cultural projects and the sustained impacts arising—including the longer-term measurement required to test these out—does appear to be limited. Where evidence is emerging, distributive effects and regeneration objectives as now defined, are generally under-achieved—or they are not sustained:

> Useful—as opposed to accurate—evaluation reports need to consider not just the impact of arts programmes on individuals, but also their effect and the extent to which it can be and is sustained on the communities in which individuals live (Newman *et al.*, 2003, p. 320).

A conclusion seems to be that the flagship and major city-centre and waterfront cultural schemes are less about regeneration than the conventional wisdom portrays them. The expectation that they will produce sustained social and distributive economic benefits alone is arguably an unreasonable one. Whether this a question of measurement, of asking the right questions, or the need for a more fundamental basic 'zero base' evaluation, is still an issue which is yet to be confronted, let alone resolved. Certainly, the nature of cultural projects which feature in regeneration schemes may need to be assessed more rigorously in terms of the impacts they actually produce—i.e. it is not only the opportunity cost *between* cultural and 'non-cultural' investment in regeneration, but between *which* type of cultural intervention and *where*, best serves the regeneration and community objectives.

Community Ownership

Capturing baseline information and building evaluation questions into project assessment is therefore essential. Attention to basic social and economic data gathering and generation will be needed if any serious attempt at measuring effects and policy evaluation is being considered. The integration of project evaluation and clearly establishing the criteria against which 'success' is measured, also need to recognise that the criteria should be set by those benefiting and participating in the cultural activity itself

> To date too little attention has been paid to the voices of ordinary citizens whose cities have been reshaped, who live with these landscapes every day and whose experiences would validate or refute the theses put forward by others (Hall, 2004, p. 71).

This sentiment is echoed by Garcia: "the emphasis must lie in providing a platform for the local communities ... to express their views and expectations" (Garcia, 2004, p. 324). This therefore echoes the need for greater community planning approaches which have been articulated in the development process in general (Healey, 1997; Solesbury, 1998).

In practice, local community involvement and the sense they might have of their 'place', is the least evident in this process, as the professional regeneration and cultural intermediaries control the territory and the rhetoric required to maintain the credibility of the expectations of culture-led regeneration. The institutional and major city cultural centres and events retain a residual and in some cases a symbolic value (Willis, 1991, p. 13), despite in many cases their declining popularity (especially in the performing arts) and narrow user-base. However, it is the everyday lived cultural practices and experiences (Lefebvre, 1991) which, the evidence suggests, better represent cultural regeneration occurring through primarily social and community-based projects.

Every Town Should Have One

In the North West region's response to the UK government's consultation on culture and regeneration (see above, DCMS, 2004), they rejected the "primary focus on new landmark investments as the route to regeneration". Rather, culture "needs to become more firmly embedded in regeneration policy and practice" (Culturenorthwest, 2004, p. 1). This is significant given that this region has hosted several major flagship regeneration projects represented by large cultural buildings and infrastructure—in Salford Quays, Manchester city centre and Northern Quarter, post-Commonwealth Games, and in a successful bid for 'Capital of Culture' in Liverpool, 2008. The Royal Institute of British Architects' annual conference held in Rotterdam, July 2003, also focused on culture-led regeneration in Liverpool. The vital issue of what was going to go in the iconic Fourth Grace building[3] prompted a comment: "If we can't decide, it won't happen" (RIBA, 2003, p. 2). In 2004, the decision was made by the city that the Fourth Grace would *not* happen. Emblematic buildings were rejected, as was the idea that culture, broadly defined, can be used to revive declining cities. During the 1990s, over 50 per cent of capital funding of culture in this entire region was accounted for in 6 projects in the 2 largest cities of Manchester and Liverpool and this concentration is mirrored in the distribution of European regional funding to visitor-led 'cultural' projects to regional cities throughout the 1990s (Evans and Foord, 2000). This reflects the economies of scale required leading to the location of major facilities within large conurbations, and also the boosterist and competitive city strategies which now look to a distinctive design statement and image.

A problem presented by this response, is the towns and cities who still aspire to cultural city status, to their own iconic projects—or 'every town should have one' (Lane, 1978)—and who look to those recent winners, literally, in the cultural lottery game. This game is played out in cities in the earlier stages of culture and regeneration, such as Singapore 'Global City of the Arts' (Chang, 2000), Adelaide (Montgomery, 2003) and Helsinki (Verwijnen and Lehto-vuori, 1999), and many others who wish to be considered part of the international cultural circuit and creative industry hubs. However, a group of artists working in and around the Thames Gateway region of South East England, which is the subject of massive house-building and regeneration plans, also recognise that

> The landmark building has become a staple element of urban regeneration, especially in industrial locations. But not every town can sustain its own Tate Modern (London), and the long term sustainability of such iconic statements is being increasingly questioned ... the iconic building as regenerative catalyst may be the wrong answer (Charrette 3, 2004, p. 3).

Whether this realisation can be accepted and alternative regeneration responses developed, here and elsewhere, remains to be seen.

There are particular issues in relation to the cultural dimension of regeneration impacts, aside from more subjective aesthetic and artistic considerations. These include the absence of planning norms for cultural facilities, against which to measure the quality and quantity of provision, and a hierarchy of such provision in terms of art form, practice and preference for the 'shock of the new' over the established, and the visible over informal and community-based culture. The latter is contrary to much of the evidence of the superior and sustained benefits of participatory arts activity compared with passive cultural consumption (a feature of most culture-led regeneration schemes). This can also reinforce the perceptual barriers to institutional spaces and places, particularly from those with lower 'cultural capital', but to whom regeneration benefits are most directed.

Economic and facility planning models do not appear to be used or successfully developed to support decision-making in this process. One consequence of the competitive 'cultural city' approach is therefore the drive for larger schemes and associated spectacular architecture, directly or indirectly, at the cost of more local and accessible and cultural provision. As Borja and Castells point out

> In practice higher layers of government replace local government through sectoral programmes or individual projects. In other cases action is taken by the private sector, without being integrated into a coherent urban programme. In yet other cases, a major area of the city and of inhabitants are simply left without any cultural facilities (Borja and Castells, 1997, p. 113).

One might conclude that the preference and trend for more social and 'cultural' (for example, heritage) evaluation is in part an admission that the ongoing economic effects from culture-led regeneration are disappointing and do not pass the 'additionality' test, but also in part, to the impact assessment models and resourcing required to measure and attribute economic impacts over the medium term.

Reasons for the Shortfall in Evidence

As summarised above in the case of the 'hierarchy' of types of information available in this field, evidence may exist but not be published or made public; or may exist in general form—via regeneration assessments—but not specifically analysed in cultural terms. More often, however, the rationale for measuring cultural impacts in relation to regeneration is not sufficiently understood or valued by stakeholders. In particular, culture is not generally recognised in urban policy or environmental and quality of life indicators (such as health, education, employment, crime) and therefore is absent from regeneration measurement criteria. From the gaps in the literature and available guidance, there is a need for a comprehensive evaluation model of a major culture-led regeneration scheme and which would serve as a practical blueprint for others.

Measuring the contribution that culture can and does make to regeneration is primarily viewed as an externality. However, the current claims for culture-led regeneration

schemes, as commonly made in their advocacy and promotion, imply that these effects
are endogenous, almost guaranteed. Nonetheless, internal barriers to the gathering of evidence of impact also exist within the cultural
sector and state funding and planning
systems. The most common include a lack
of interest on the part of the cultural sector
in developing evaluative systems through
which to prove its value and the view held
by some creative practitioners in particular,
that evaluation is an unnecessary, bureaucratic
intrusion in the creative process (Matarasso,
1996, p. 24).

The reasons for the barriers and resistance
to the evaluation of impacts are therefore 'cultural' on the one hand and structural on the
other, including the rationale for the resources
needed to undertake the required gathering of
evidence at the outset and over time. Today,
few would dispute the role and value that
culture has in regeneration in the narrow
and, increasingly, in the wider sense, but
there is much less understanding of the very
different effects that different types of cultural
intervention produce in the short and longer
term

> Culture-led regeneration does change
> people's lives after all. It is about time we
> understood how and why it does so
> (Bailey et al., 2004, p. 64).

Despite a mixed experience of flagship and
iconic buildings and mega-events, and the
regenerative effects of these costly grands
projets, these formulas and strategies continue
to be emulated, whether in making the case for
a Guggenheim franchise in Rio or Liverpool,
in revisioning a post-industrial city, or in
bidding for international events and festival
sites. More grounded evidence and assessment of the cultural opportunities is therefore
needed, as much as of general regeneration
programme outcomes. For practitioners,
researchers, community groups and policy-
makers, developing an appropriate evaluation
model and schema from the set of indicators
and principles now available, is recommended. This requires learning selectively from
the 'evidence' which must be conditioned by

what is an unhelpful but endemic bias in this
field. A pluralist rather than a standardised
approach is therefore an imperative, since
this is unlikely to emerge from the regeneration regimes and 'evidence base' currently
on offer

> Culture . . . can make communities. It can be
> a critical focus for effective and sustainable
> urban regeneration. The task is to develop an
> understanding (including methods of study)
> of the ways—cultural and ethical—in which
> even the 'worst estates' can take part in and
> help shape the relics of their city (and
> society) as well as their locality. This is a
> massive challenge to academics, pro-
> fessionals, business, and to local and ulti-
> mately national government and—of
> course—citizens. But nothing less can
> work (Catterall, 1998, p. 4).

Notes

1. The review undertaken by the author was
    commissioned by the UK Department for
    Culture, Media and Sport (DCMS) in 2003–
    04. It comprised a critical literature review
    of 'published evidence' and case studies of
    culture's contribution to regeneration and
    good practice in evaluation. The definition
    of cultural activity used by the DCMS for
    this review includes the arts (including film),
    libraries, museums, heritage and cultural
    tourism. For the scope and methodology
    used, see p. 3 of the review. The DCMS
    consultation report arising: Culture at the
    heart of regeneration was launched by the
    Secretary of State in June 2004. This report
    and the review are both available at
    www.culture.gov.uk.

2. For a detailed review and bibliography of
    impact and evaluation studies in the culture
    and regeneration field, and used for this
    paper, see Evans and Shaw (2004) (www.
    culture.gov.uk) and, on social impacts, see
    Evans and Shaw (2001a, 2001b) (www.
    citiesinstitute.org).

3. The 'Fourth Grace', named 'The Cloud' by its
    architect Will Alsop, was designed to comp-
    lement Liverpool's 'Three Graces'—the
    Royal Liver Building, the Cunard Building
    and the Port of Liverpool Building—on Mer-
    seyside's Pier Head. This spectacular flagship
    formed a key part of Liverpool's successful
    bid for 'European Capital of Culture 2008':

"a focus and catalyst for the next stage of Liverpool's renaissance, an eloquent image for a resurgent city" (www.liverpoolfourth-grace.co.uk). Following the 'Capital of Culture' designation, the City Council cancelled this project, citing 'spiralling costs' (forecast to rise from £228 million to £324 million), unclear usage and the experience of 'out of control' iconic building projects in other places.

References

ANDRA, I. (1987) The dialectic of tradition and progress, in: *Architecture and Society: In Search of Context*, pp. 156–158. Sofia: Balkan State Publishing House.

ARTS COUNCIL (1999) *Best value and the arts. Practical guidance to Assist Local Authority Arts Officers and Others in Applying the Duty of 'Best Value' to The Arts*. London: Arts Council of England.

BAILEY, C., MILES, S. and STARK, P. (2004) Culture-led urban regeneration and the revitalisation of identities in Newcastle, Gateshead and the North East of England, *International Journal of Cultural Policy*, 10(10), pp. 47–65.

BALIBREA, M. P. (2001) Urbanism, culture and the post-industrial city: challenging the Barcelona model, *Journal of Spanish Cultural Studies*, 2(2), pp. 187–210.

BANIOTOPOULOU, E. (2000) Art for whose sake? Modern art museums and their role in transforming societies: the case of the Guggenheim Bilbao, *Journal of Conservation and Museum Studies*, 7, pp. 1–15.

BASSETT, K. (1993) Urban cultural strategies and urban regeneration: a case study and critique, *Environment & Planning A*, 25(12), pp. 1773–1789.

BASSETT, K., GRIFFITHS, R. and SMITH, I. (2002) Testing governance: partnerships, planning and conflict in waterfront regeneration, *Urban Studies*, 39(10), pp. 1757–1775.

BCC (BIRMINGHAM CITY COUNCIL) (2003) *Birmingham's Renaissance: How European Funding has Revitalised the City*. Birmingham: Birmingham City Council.

BENNETT, T. (1998) The multiplication of culture's utility, *Critical Inquiry*, 21, pp. 861–889.

BETTERTON, J. (2001) Culture and urban regeneration: the case of Sheffield, *Sub-Regional Commentary*, S.Yorkshire, p. 11.

BIANCHINI, F. (1991) Alternative cities, *Marxism Today*, June, pp. 36–38.

BIANCHINI, F. and GHILARDI, L. (1997) *Culture and Neighbourhoods, Vol. 2. A Comparative Report*. Strasbourg: Council of Europe.

BODDY, M. and PARKINSON, M. (2004) *City Matters: Competitiveness, Cohesion and Urban Governance*. Bristol: Policy Press.

BODO, C., GORDON, C. and ILCZUK, D. (Eds) (2004) *Gambling on Culture: State Lotteries as a Source of Funding for Culture—The Arts and Heritage*. Amsterdam: CIRCLE/Boekmanstudies.

BOGNER, D. (2001) A product of chance or planned diversity? The concept of the MuseumsQuartiers, in: M. BOECKL (Ed.) *MuseumsQuartier Wien*, pp. 33–40. Vienna: Springer-Verlag.

BORDAGE, F. (Ed.) (2002) *The Factories: Conversions for Urban Culture*. Basel: Birkhauser.

BORJA, J. and CASTELLS, M. (1997) *Local and Global: Management of Cities in the Information Age*. London: Earthscan.

BOWLING, A. (1997) *Measuring Health: A Review of Quality of Life Measurement Scales*. Milton Keynes: Open University Press.

CABE (2002) *Design Quality Indicators*. London: Construction Industry Council.

CARMONA, M., MAGALHAES, C. DE, and EDWARDS, M. (2002) Stakeholder views on value and urban design, *Journal of Urban Design*, 7(2), pp. 145–169.

CATTERALL, B. (1998) *Culture as a critical focus for effective urban regeneration*. Town & Country Planning Summer School, University of York.

CHANG, T. C. (2000) Renaissance revisited: Singapore as a "Global City for the Arts", *International Journal of Urban and Regional Research*, 24(4), pp. 818–831.

CHARRETTE 3 (2004) *New models of cultural facilities*. Tilbury: Thurrock, 26 May.

CLRAE (1992) European Urban Charter. Adopted by the Council of Europe's Standing Conference of Local and Regional Authorities of Europe on 18th March. Strasbourg.

CONNOLLY, S. (1997) The measurement of additionality: a case study of the UK National Lottery, in: *Business and Economics in the 21st Century, Vol. 1*, BES International Conference.

COOTE, A., ALLEN, J. and WOODHEAD, D. (2004) *Finding Out What Works: Building Knowledge about Complex, Community-based Initiatives*. London: King's Fund.

CRAWFORD, L. (2001) Bilbao thrives from the "Guggenheim effect", *Financial Times Weekend*, 28 April, p. 2.

CULTURENORTHWEST (2004) *Culture at the Heart of Regeneration*. Manchester, 30 September (www.culturenorthwest.co.uk).

DCMS (DEPARTMENT FOR CULTURE, MEDIA AND SPORT) (1999) *Draft Guidance for Local Cultural Strategies*. London: DCMS.

DCMS (2000) *The White Book—Option Appraisal of Expenditure Decisions. A Guide for the DCMS and its Sponsored Bodies*. London: DCMS.

DCMS (2003) *Building tomorrow: culture in regeneration.* The Lowry, Manchester, 25 February (www.culture.gov.uk).

DCMS (2004) *Culture at the heart of regeneration.* London: DCMS.

DETR (DEPARTMENT FOR THE ENVIRONMENT, TRANSPORT AND THE REGIONS) (1998) *Sustainable development indicators—local quality of life counts.* London: DETR.

DTI (DEPARTMENT OF TRADE AND INDUSTRY) (2004) *Creative people, openness and productivity: an exploration of regional differences Inspired by 'The Rise of the Creative Class'.* London: DTI.

DUDRAH, R. K. (2002) Birmingham (UK): constructing city spaces through Black popular cultures and the Black public sphere, *City*, 6(3), pp. 335–350.

EC (EUROPEAN COMMISSION) (1998) *Social Indicators: Problematic Issues.* Brussels: EC.

EHRENREICH, B. and EHRENREICH, J. (1979) The professional-managerial class, in: P. WALKER (Ed.) *Between Labour and Capital*, pp. 5–45. Sussex: Harvester.

EVANS, G. L. (1998) Study into the employment effects of arts lottery spending in England. Research Report No. 14, Arts Council, London.

EVANS, G. L. (2000) Measure for measure: evaluating performance and the arts organisation, *Studies in Cultures, Organizations and Societies*, 6(2), pp. 243–266.

EVANS, G. L. (2001a) *Cultural Planning: An Urban Renaissance?* London: Routledge.

EVANS, G. L. (2001b) The World Bank and world heritage: culture and sustainable development?, *Tourism Recreation Research*, 26(3), pp. 83–86.

EVANS, G. L. (2003) Hard-branding the cultural city: from Prado to Prada, *International Journal of Urban and Regional Research*, 27(2), pp. 417–440.

EVANS, G. L. (2004a) Cultural industry quarters—from pre-industrial to post-industrial production, in: D. BELL and M. JAYNE (Eds) *City of Quarters: Urban Villages in the Contemporary City*, pp. 71–92. Aldershot: Ashgate.

EVANS, G. L. (2004b) Measuring impact: culture and regeneration, *Arts Professional*, 83(November), pp. 5–6.

EVANS, G. L. and FOORD, J. (2000) European funding of culture: promoting common culture or regional growth?, *Cultural Trends*, 36, pp. 53–87.

EVANS, G. L. and FOORD, J. (2003) Cultural planning in east London, in: N. KIRKHAM and M. MILES (Eds) *Cultures and Settlement: Advances in Art and Urban Futures, Vol. 3*, pp. 15–30. Bristol: Intellect Books.

EVANS, G. L. and SHAW, P. (2001a) *Study into the Social Impact of Lottery Good Cause Spending.* London: Department for Culture Media and Sport.

EVANS, G. L. and SHAW, S. (2001b) Urban leisure and transport: regeneration effects, *Journal of Leisure Property*, 1/4, pp. 350–372.

EVANS, G. L. and SHAW, P. (2004) *A review of evidence on the role of culture in regeneration.* London: Department for Culture Media and Sport (www.culture.gov.uk).

FCM (FEDERATION OF CANADIAN MUNICIPALITIES) (2001) *The FCM quality of life reporting system: second report quality of life in Canadian cities.* March. Ottawa: FCM.

FEIST, A., FISHER, R., GORDON, C. and MORGAN, C. (1998) International data on public spending on the arts in eleven countries. Research Report No. 13, Arts Council of England (ACE), London.

FLORIDA, R. (2004) *The Rise of the Creative Class*, 2nd edn. New York: Basic Books.

FORUM BARCELONA 2004 (2000) *Agenda of principles and values: universal forum of cultures* (www.barcelona2004.org).

GARCIA, B. (2004) Cultural policy and urban regeneration in western European cities: lessons from experience, prospects for the future, *Local Economy*, 19(4), pp. 312–326.

GDANIEC, C. (2000) Cultural industries, information technology and the regeneration of post-industrial urban landscapes. Poblenou in Barcelona—a virtual city?, *Geojournal*, 50(4), pp. 379–388.

GEHL, J. (2001) *Life between Buildings: Using Public Spaces*, 4th edn. Denmark: Arkitektens Forlag.

GERTLER, M., FLORIDA, R., GATES, G. and VINODRAI, T. (2002) *Competing on creativity: placing Ontario's cities in North American context.* Toronto: Ontario Ministry of Enterprise, Opportunity and Innovation.

GILMORE, F. (2004) Shanghai: unleashing creative potential, *Journal of Brand Management*, 11(6), pp. 442–448.

GRATTON, C. and TAYLOR, P. (1995) Impacts of festival events: a case study of Edinburgh, in: G. ASHWORTH and A. DIETWVORST (Eds) *Tourism and Spatial Transformations*, pp. 225–238. Wallingford: CAB International.

GRIFFITHS, R. (1993) The politics of cultural policy in urban regeneration strategies, *Policy and Politics*, 87, pp. 84–94.

HALL, C. M. (1992) *Hallmark Tourist Events: Impacts, Management, Planning.* London: Belhaven.

HALL, P. (1998) *Cities and Civilization: Culture, Innovation, and Urban Order.* London: Weidenfeld & Nicholson.

HALL, T. (2004) Public art, civic identity and the new Birmingham, in: L. KENNEDY (Ed.) *Remaking Birmingham*, pp. 63–71. London: Routledge.

HALL, T. and ROBERTSON, I. (2001) Public art and urban regeneration: advocacy, claims and critical debates, *Landscape Research*, 26(1), pp. 5–26.

HAMILTON, J., FORSYTH, L. and IONGH, D. DE (2001) Public art: a local authority perspective, *Journal of Urban Design*, 6(3), pp. 283–296.

HANNIGAN, J. (Ed.) (2003) Symposium on branding, the entertainment economy and urban place building, *International Journal of Urban and Regional Research*, 27(20), pp. 352–360.

HARRISON, T. (2000) Urban policy: addressing wicked problems, in: H. DAVIES, S. NUTLEY and P. SMITH (Eds) *What Works: Evidence-based Policy and Practice in Public Services*, pp. 207–228. Bristol: Policy Press.

HARVEY, D. (1993) Goodbye to all that? Thoughts on the social and intellectual condition of contemporary Britain, *Regenerating Cities*, 5, pp. 11–16.

HAUBEN, T., VERMEULEN, M. and PATTEEUW, V. (Eds) (2002) *City Branding: Image Building and Building Images*. Rotterdam: NAI Uitgevers.

HAYDEN, D. (1995) *The Power of Place: Urban Landscapes as Public History*. Cambridge, MA: MIT Press.

HEALEY, P. (1997) *Collaborative Planning: Shaping Places in Fragmented Societies*. London: Macmillan.

HMSO (HER MAJESTY'S STATIONERY OFFICE) (1993) *Treaty on European Union*, Maastricht, 7 February 1992 (entered into force 1 November 1993). London: HMSO.

HUGHES, R. (1998) *Culture Makes Communities Conference*. Leeds: Joseph Rowntree Foundation, 13 February.

JACKSON, A. (2000) *Social Impact Study of the Millennium Awards*. London: The Millennium Commission.

JACKSON, A. (2005) *The Evaluation Toolkit*. Belfast: Arts Council of Northern Ireland.

JEANNOTTE, M. S. and STANLEY, D. (2002) How will we live together?, *Canadian Journal of Communication*, 27(2/3), pp. 133–141.

KAVARATZIS, M. (2004) From city marketing to city branding: towards a theoretical framework for developing city brands, *Journal of Place Branding*, 1(1), pp. 58–73.

KEARNS, G. and PHILO, C. (1993) *The City as Cultural Capital: Past and Present*. Oxford: Pergamon.

KLUNZMAN, K. (2004) Keynote speech to Intereg III Mid-term Conference, Lille, in: *Regeneration and Renewal*, 19 November, p. 2.

KPMG (1998) Impacto de las actividades de la Guggenheim-Bilbao Museoaren Fundazioa en Euskadi, Bilbao, Spain, *Connaissance des Arts*, 559, pp. 106–107.

KROLLER, E.-M. (1996) *EXPO 67: Canada's Camelot?*, in: *Proceedings of the British Association for Canadian Studies (BACS) Annual Conference*, April, Exeter University.

LANDRY, C. (2000) *The Creative City: A Toolkit for Urban Innovators*. London: Earthscan.

LANDRY, C. and BIANCHINI, F. (1995) *The Creative City*. London: DEMOS.

LANE, J. (1978) *Arts Centres—Every Town Should Have One*. London: Paul Elek.

LASH, S. and URRY, J. (1994) *Economies of Signs and Spaces*. London: Sage.

LDA (LONDON DEVELOPMENT AGENCY) (2003) *Economic Impact Study of the Notting Hill Carnival*. London: LDA.

LDA (2004) *Strategy for creative projects*. International Initiatives Unit, LDA.

LEADBEATER, C. (2000) *Living on Thin Air*. London: Penguin Books.

LEFEBVRE, H. (1974) *The Production of Space*, trans. by D. Nicholson-Smith. Oxford: Blackwell.

LEFEBVRE, H. (1991) *Critique of Everyday Life*, trans. by J. Moore. London: Verso.

LGA (LOCAL GOVERNMENT ASSOCIATION) (2000) *A change of scene: the challenge of tourism in regeneration*. London: LGA/DCMS.

LIM, H. (1993) Cultural strategies for revitalizing the city: a review and evaluation, *Regional Studies*, 27(6), pp. 589–594.

LOFTMAN, P. (1990) *A tale of two cities: Birmingham the convention and unequal city*. Research Paper No. 6, Faculty of the Built Environment, Birmingham Polytechnic.

LPAC (LONDON PLANNING ADVISORY COMMITTEE) (1990) *Strategic planning policies for the arts, culture and entertainment*. Report No. 18/90, LPAC.

MACCLANCY, J. (1997) The museum as a site of contest: the Bilbao Guggenheim, *Focal Journal of Anthropology*, 1(7), pp. 271–278.

MARSHALL, R. (Ed.) (2001) *Waterfronts in Post-industrial Cities*. London: Routledge.

MATARASSO, F. (1996) *Defining values: evaluating arts programmes, the social impact of the arts*. Working Paper 1, Comedia, Stroud.

MATARASSO, F. (1997) *Use or ornament? The social impact of participation in the arts*. Stroud: Comedia.

MCCARTHY, J. (2002) Entertainment-led regeneration: the case of Detroit, *Cities*, 19(2), pp. 105–111.

MCGUIGAN, J. (1996) *Culture and the Public Sphere*. London: Routledge.

MILES, M. (2004) Drawn and quartered: El Raval and the Haussmannization of Barcelona, in: D. BELL and M. JAYNE (Eds) *City of Quarters: Urban Villages in the Contemporary City*, pp. 37–55. Aldershot: Ashgate.

MOKRE, M. (1998) EU cultural intervention in area regeneration processes. UACES European Cultural Policy Conference, City University, London, April.

MONTGOMERY, J. (2003) *Cultural Quarters as Mechanisms for Urban Regeneration: A Review.* Adelaide: Planning Institute of Australia National Congress.

NATIONAL BUILDING MUSEUM (1998) *Building Culture Downtown: New Ways of Revitalising the American City.* Washington, DC: National Building Museum.

NEGRIER, E. (1993) Montpellier: international competition and community access, in: F. BIANCHINI and M. PARKINSON (Eds) *Cultural Policy and Urban Regeneration: The West European Experience*, pp. 235–254. Manchester: Manchester University Press.

NEWMAN, A. and MCLEAN, F. (1998) Heritage builds communities: the application of heritage resources to the problems of social exclusion, *International Journal of Heritage Studies*, 4, pp. 143–153.

NEWMAN, T., CURTIS, K. and STEPHENS, J. (2003) Do community-based arts projects result in social gains? A review of the literature, *Community Development Journal*, October, pp. 310–322.

NICHOLS CLARK, T. (Ed.) (2004) *The City as Entertainment Machine.* Oxford: Elsevier.

ODPM (OFFICE OF THE DEPUTY PRIME MINISTER) (2001) *Towns & cities: partners in urban renaissance. Breaking down the barriers.* London: ODPM/CABE.

ODPM (2003) *Assessing the impacts of spatial interventions. Regeneration, renewal and regional development. Main guidance* (Consultation, April). London: ODPM.

PADDISON, R. (1993) City marketing, image reconstruction and urban regeneration, *Urban Studies*, 30(2), pp. 339–350.

PALMER, R. (2004) *European capitals/cities of culture: study on the European cities and capitals of culture and the European cultural months (1995–2004)*, Part I and II. Palmer/Rae Associates. Brussels: European Commission.

PERRI 6 (1997) *Holistic Government.* London: DEMOS.

PLAZA, B. (2000) Evaluating the influence of a large cultural artifact in the attraction of tourism: the Guggenheim Museum Bilbao case, *Urban Affairs Review*, 36(2), pp. 264–274.

PMSU (PRIME MINISTER'S STRATEGY UNIT) (2004) *Strategy Survival Guide.* London: PMSU (www.strategy.gov.uk/su/survivalguide/index.htm).

PORTER, M. E. (1995) The competitive advantage of the inner city, *Harvard Business Review*, May/June, pp. 55–71.

PRENTICE, R. and ANDERSEN, V. (2003) Festival as creative destination, *Annals of Tourism Research*, 30(1), pp. 7–30.

RAPLEY, M. (2001) *Quality of Life Research.* London: Sage.

RIBA (ROYAL INSTITUTE OF BRITISH ARCHITECTS) (2003) *Annual Conference Rotterdam 2003: Regeneration.* London: RIBA.

ROBERTS, N. and MARSH, C. (1995) For art's sake: public art, planning policies and the benefits for commercial property, *Planning Practice and Research*, 2, pp. 189–198.

ROBINS, K. (1996) Collective emotion and urban culture, in: B. BRANDER, S. MATTLE and V. RATZENBOK (Eds) *Kulturpolitik und Restrukturierung der Stadt*, pp. 73–96. Vienna: International Archives for Culture Analyses.

RODRIGUEZ, A., MARTINEZ, E. and GUENGA, G. (2001) Uneven development: new urban policies and socio-spatial fragmentation in metropolitan Bilbao, *European Urban and Regional Studies*, 8(2), pp. 161–178.

RUIZ, J. (2004) *A literature review of the evidence base for culture, the arts and sport policy.* Scottish Executive Education Department, Edinburgh, February.

SCALBERT, R. (1994) Have the *Grands Projets* really benefited Paris?, *Architect's Journal*, 3(200), p. 20.

SELWOOD, S. (1995) *The Benefits of Public Art.* London: Policy Studies Institute.

SENNETT, R. (1986) *The Fall of Public Man.* London: Faber & Faber.

SEO, J.-K. (2002) Re-urbanisation in regenerated areas of Manchester and Glasgow–new residents and the Problems of Sustainability, *Cities*, 19(2), pp. 113–121.

SHAW, P. (1990) *The public art report: commissions by local authorities.* London: Public Arts Development Trust.

SHAW, P. (1999) *The arts and neighbourhood renewal: a research report.* Policy Action Team 10. London: Department for Culture, Media and Sport.

SHAW, P. (2003) *What's art got to do with it? Briefing paper on the role of the arts in neighbourhood renewal.* London: Arts Council of England.

SIMMIE, J. (Ed.) (2001) *Innovative Cities.* London: Spon Press.

SOLESBURY, W. (1998) *Good connections: helping people to communicate in cities.* Working paper No. 9, Comedia, Stroud.

STEVENSON, D. (2004) 'Civic gold' rush: cultural planning and the politics of the third way, *International Journal of Cultural Policy*, 10(1), pp. 119–131.

STOKER, G. (1995) Regime theory and urban politics, in: D. JUDGE, G. STOKER and H. WOLMAN

(Eds) *Theories of Urban Politics*, pp. 54–71. London: Sage.

SUDJIC, D. (1993) *The 100 Mile City*. London: Flamingo.

SYMON, P. (1999) *The new arts in Birmingham: a local analysis of cultural diversity*, in: *Proceedings of the International Conference on Cultural Policy Research*, pp. 723–744. Bergen, November.

THOMAS, C. J. and BROMLEY, D. F. (2000) City-centre revitalisation: problems of fragmentation and fear in the evening and night-time city, *Urban Studies*, 37(8), pp. 1403–1429.

TREASURY, HM (2003) *Appraisal and Evaluation in Central Government* (The Green Book). London: HM Treasaury.

ULLDEMOLINS, J. R. (2000) From "Chino" to Raval: art merchants and the creation of a cultural quarter in Barcelona, in: *Proceedings of the Cultural Change and Urban Contexts Conference*, p. 19. Manchester, September.

VERWIJNEN, J. and LEHTOVUORI, P. (Eds) (1999) *Creative Cities: Cultural Industries, Urban Development and the Information Society*. Helsinki: University of Art and Design Press.

WANG, C. (2002) *Waterfront regeneration*. Town & Country Planning Summer School, Cardiff University, Wales.

WARD, S. (1998) *Selling Places: The Marketing and Promotion of Towns and Cities 1850–2000*. London: E & FN Spon.

WARD THOMSON, C., PATRIZIO, A. and MONTARZINO, A. (2004) *Research on public art: assessing impact and quality* (draft report). Edinburgh College of Art, October.

WILLIAMS, D. (1996) *Creating social capital: a study of the long-term benefits from community based arts funding*. Community Arts Network of South Australia/Australia Council of the Arts.

WILLIS, P. (1991) *Towards a new cultural map*. Discussion Document No. 18, National Arts & Media Strategy. Arts Council, London.

WORPOLE, K. and GREENHALGH, L. (1999) *The richness of cities: urban policy in a new landscape—final report*. Stroud: Comedia/Demos.

WU, F. (2000) The global and local dimensions of place-making: remaking Shanghai as a world city, *Urban Studies*, 37, pp. 1359–1377.

Sport and Economic Regeneration in Cities

Chris Gratton, Simon Shibli and Richard Coleman

Introduction

In the UK, in the 1970s and early 1980s, government expenditure on sport expanded considerably. The rationale for this increased expenditure was that sport made a considerable contribution to local communities in welfare terms. Following the publication of the White Paper *Sport and Recreation* (Department of the Environment, 1975), it was established that sport should be regarded as part of the general fabric of the social services. Most of this additional expenditure was made by local government on indoor sports centres and swimming pools. In 1971, there were 12 indoor sports centres and 440 swimming pools in Britain. By 1981, there were 461 indoor sports centres and 964 swimming pools (Gratton and Taylor, 1991). This growth in expenditure came to an end in the mid 1980s with the public expenditure cuts of the then Conservative government.

At the same time as the investment in sport for welfare reasons started to decline, a second wave of sports investment began, but this time the rationale was economic regeneration. Investment in sport infrastructure in cities was not primarily aimed at getting the local community involved in sport, but was instead aimed at attracting tourists, encouraging inward investment and changing the image of the city. The first example of this new strategy was seen in Sheffield with the investment of £147 million in sporting facilities to host the World Student Games of 1991. There were also the Olympic bids of Birmingham and Manchester in the 1980s and 1990s. These did not immediately result in investment in facilities since the bids were unsuccessful, but substantial expenditure was required just to mount the bids. More recently, Manchester spent over £200 million on sporting venues in order to host the 2002 Commonwealth Games, with a further £470 million expenditure on other non-sport infrastructure investment in Sportcity in east Manchester.

In the British context, most of the cities following this strategy of using sport for economic regeneration were industrial cities, not normally known as major tourist destinations. The drivers of such policies were the need for a new image and new employment opportunities caused by the loss of their conventional industrial base. In the US, cities such as Indianapolis and Cleveland had adopted a similar strategy in the 1970s and 1980s, again following increased unemployment due to deindustrialisation. However, in the US, sport-related regeneration strategies have tended to be focused on facilities for domestic professional team sports rather than on hosting major international sports events. In the rest of Europe and Australia, we have seen similar strategies—most notably in Barcelona with the hosting of the 1992 Olympics, in Athens with the 2004 Olympics and in Sydney with the 2000 Olympics. The differences between these cities and the British and American ones are that they are already major tourist destinations in their own right prior to hosting the Olympics and were not facing the same problems of industrial decline. The objective here was to transform the image of these cities and turn them into major world cities.

In this article, we analyse the justification for such investments in sport in cities and assess the evidence on the success of such strategies.

Sport and Urban Regeneration

The study of hallmark events or mega-events became an important part of tourism literature in the 1980s. Since then, the economics of sports tourism at major sports events has become an increasing part of this event tourism literature. Many governments around the world have adopted national sports policies that specify that hosting major sports

events is a major objective. A broad range of benefits has been suggested for both the country and the host city including: urban regeneration legacy benefits, sporting legacy benefits, tourism and image benefits, social and cultural benefits and the direct economic impact benefits which will be the main focus of this article. It is well known that cities and countries compete fiercely to host the Olympic Games or the soccer World Cup. However, over recent years there has been increasing competition to host less globally recognised sports events in a wide range of other sports where spectator interest is less certain and where the economic benefits are not so clearcut. In this review, we will analyse the benefits generated across a wide range of sports events from large spectator events staged as part of domestic professional team sports to World and European Championships. We will concentrate on the economic benefits generated, but will also consider the broader benefits outlined above. To begin, we discuss the literature associated with hosting major sports events.

The literature on the economics of major sports events is relatively recent. One of the first major studies in this area was the study of the impact of the 1985 Adelaide Formula 1 Grand Prix (Burns et al., 1986). This was followed by Brent Richie's in-depth study of the 1988 Calgary Winter Olympics (Richie, 1984; Richie and Aitken, 1984, 1985; Richie and Lyons, 1987, 1990; Richie and Smith, 1991). In fact, immediately prior to these studies it was generally thought that hosting major sports events was a financial liability to host cities following the large debts faced by Montreal after hosting the 1976 Olympics. There was a general change in attitude following the 1984 Los Angeles Olympics which made a clear profit.

Mules and Faulkner (1996) point out that even with such mega-events as Formula 1 Grand Prix races and the Olympics, it is not always an unequivocal economic benefit to the cities that host the event. They emphasise that, in general, staging major sports events often results in the city authorities losing money even though the city itself benefits greatly in terms of additional spending in the city. Thus the 1994 Brisbane World Masters Games cost the city A$2.8 million to organise but generated a massive A$50.6 million of additional economic activity in the state economy. Mules and Faulkner's basic point is that it normally requires the public sector to finance the staging of the event and incurring these losses in order to generate the benefits to the local economy. They argue that governments host such events and lose taxpayers' money in the process in order to generate spillover effects or externalities.

It is not a straightforward job, however, to establish a profit and loss account for a specific event. Major sports events require investment in new sports facilities and often this is paid for in part by central government or even international sports bodies. Thus some of this investment expenditure represents a net addition to the local economy since the money comes in from outside. Also, such facilities remain after the event has finished acting as a platform for future activities that can generate additional tourist expenditure (Mules and Faulkner, 1996).

Sports events are increasingly seen as part of a broader tourism strategy aimed at raising the profile of a city and therefore success cannot be judged on simply profit and loss basics. Often the attraction of events is linked to a reimaging process and, in the case of many cities, is invariably linked to strategies of urban regeneration and tourism development (Bianchini and Schwengel, 1991; Bramwell, 1995; Loftman and Spirou, 1996; Roche, 1994). Major events if successful have the ability to project a new image and identity for a city. The hosting of major sports events is often justified by the host city in terms of long-term economic and social consequences, directly or indirectly resulting from the staging of the event (Mules and Faulkner, 1996). These effects are primarily justified in economic terms, by estimating the additional expenditure generated in the local economy as the result of the event, in terms of the benefits injected from tourism-related activity and the subsequent reimaging of the city following the success of the event (Roche, 1992).

Cities staging major sports events have a unique opportunity to market themselves to the world. Increasing competition between broadcasters to secure broadcasting rights to major sports events has led to a massive escalation in fees for such rights, which in turn means that broadcasters give blanket coverage of such events at peak times, enhancing the marketing benefits to the cities that stage them.

Such benefits might include a notional value of exposure achieved from media coverage and the associated place marketing effects related to hosting and broadcasting an event which might encourage visitors to return in future or, alternatively, an investigation into any sports development impacts, which may encourage young people to get more involved in sport. Collectively, these additional benefits could be monitored using a more holistic 'balanced scorecard' approach to event evaluation as outlined in Figure 1.

In theory, then, there is a wide diversity in the range of economic benefits that sports events can generate. Kasimati (2003) summarised the potential long-term benefits to a city of hosting major sports events such as the summer Olympics: newly constructed event facilities and infrastructure, urban revival, enhanced international reputation, increased tourism, improved public welfare, additional employment and increased inward investment. In practice, however, there is also a possible downside to hosting such events including: high construction costs of sporting venues and related other investments in particular in transport infrastructure, temporary congestion problems, displacement of other tourists due to the event and underutilised élite sporting facilities after the event which are of little use to the local population.

Gibson (1998) indicates that there is a growing body of literature which adopts a more critical approach towards the impacts of events suggesting that the negative impacts may outweigh the benefits. Roche (1994) investigated the policy processes that led to the hosting of the World Student Games in Sheffield in 1991. He argued that there was no rational policy process, no

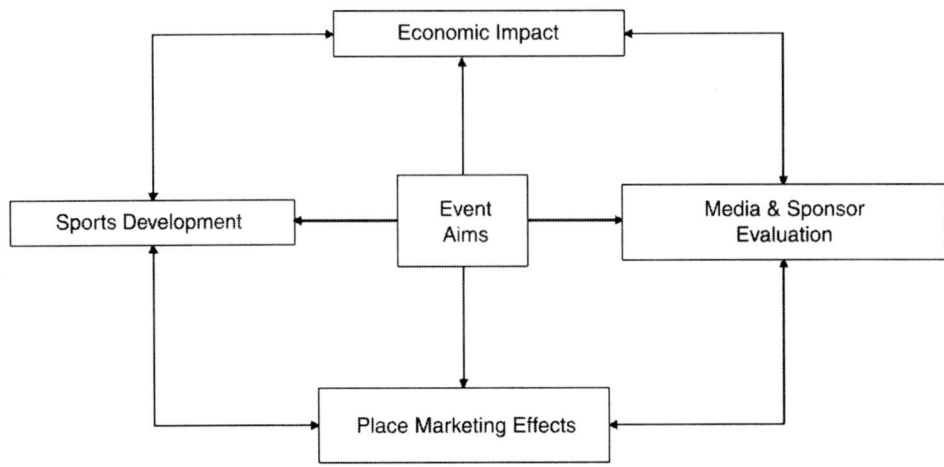

Figure 1. The 'balanced scorecard' approach to evaluating events.

effective tourism strategy, nor any organisational means to implement such a strategy. He argued that the decision-making processes that led to the successful bid for the event were non-rational and non-democratic.

Sack and Johnson (1996) conducted a similar study into the hosting of the Volvo International Tennis Tournament in New Haven Connecticut in 1989. Their conclusions are in agreement with Roche that all the main policy decisions were made without consultation with the general public and the new tennis stadium built for the event at a cost of $15 million was not open to the general public except when special events were held.

Roche (1994) argues that decisions about hosting major events are essentially political decisions made by urban political leaderships and/or other relevant and powerful urban élite groups with little or no democratic community input. The justification for the event may be put in economic terms, but the evidence on the economic benefits is rarely available at the time the decision is made.

Kasimati (2003) analysed all impact studies of the summer Olympics from 1984 to 2004 and found, in each case, that the studies were done prior to the Games, were not based on primary data and were, in general, commissioned by proponents of the Games. He found that the economic impacts were likely to be inflated since the studies did not

take into account supply-side constraints such as investment crowding out, price increases due to resource scarcity and the displacement of tourists who would have been in the host city had the Olympics not been held there.

Thus, despite a strong theoretical case in favour of urban regeneration benefits from investment in sporting infrastructure in order to host major sports events, there are also strong arguments that the negative impacts of such investment may match or even outweigh these benefits. We now turn to the evidence.

City Sports Strategies in North America

Over the past two decades, many cities in the US have invested vast amounts of money on sports stadia on the basis of arguments relating to economic benefits to the city from such investment. Most of these strategies have been based on professional team sports, in particular, American football, baseball, ice-hockey and basketball. Unlike the situation in Europe, professional teams in North America frequently move from city to city.

Since the late 1980s, cities have offered greater and greater incentives for these professional teams to move by offering to build new stadia to house them costing hundreds of millions of dollars. The teams just sit back and let cities bid up the price. They either move to the city offering the best deal or they accept the counter offer invariably put to them by their existing hosts. This normally involves the host city building them a brand new stadium to replace their existing one which may only be 10 or 15 years old.

Baade indicates how, since the 1980s, escalating stadium construction costs have increased the size of stadium subsidies

> The number of stadiums that have been built since 1987 to the present is unprecedented. Approximately 80 per cent of the professional sports facilities in the United States will have been replaced or have undergone major renovation during this period of time. The new facilities have cost more than $19 billion in total, and the public has provided $13.6 billion, or 71 per cent, of that amount. In few, if any, instances have professional teams in the United States been required to open their books to justify the need for these subsidies. Rather, teams have convinced cities that to remain competitive on the field they have to be competitive financially, and this, teams claim, cannot be achieved without new playing venues (Baade, 2003, p. 588).

This use of taxpayers' money to subsidise profit-making professional sports teams seems out of place in the North American context. The justification for such public expenditure is an economic one: the investment of public money is a worthwhile investment as long as the economic impact generated by having a major professional sports team resident in the city is sufficiently great. Economic impact refers to the total amount of additional expenditure generated within a host city (or area) which could be directly attributable to the staging of a particular event. Only visitors to the host economy as a direct result of an event being staged are eligible for inclusion in the economic impact calculations (i.e. the expenditure by people resident in the host area is not included on the basis that they would spend money locally irrespective of whether an event is taking place).

Baade (1996), Noll and Zimbalist (1997) and Coates and Humphreys (1999), however, showed no significant direct economic impact on the host cities from such stadium development. Crompton (1995, 2001) also argues that economic impact arguments in favour of such stadium construction using public subsidies have been substantially exaggerated. However, he goes on to suggest (Crompton, 2001, 2004) that there are other possible benefits to cities from such developments: increased community visibility, enhanced community image, stimulation of additional development related to the stadium and psychic income to city residents from having a professional team in the city. The first three of these focus on the ability of such stadium developments to influence external audiences which may lead to inward investment into the host city and generate similar benefits to economic impact. Psychic income relates to the social and psychological benefit local residents may feel by identifying with the resident professional team. Although sports researchers are well aware of such benefits, they are notoriously difficult to measure effectively and no evidence currently exists to suggest that these broader benefits justify the high levels of public subsidies to professional sports teams in the US.

The question that arises, therefore, is why such subsidies have grown to these massive levels in recent years. Quirk and Fort suggest an answer to this question

> As monopolies, sports leagues artificially restrict the number of teams below the number that would be in business if there was competition in the sport. By constantly keeping a supply of possible host cities—cities that could support a league team—on line, current host cities are in the unenviable position of being pressured to provide exorbitant subsidies to their teams or risk losing them (Quirk and Fort, 1999, pp. 169–170).

Thus it is simply a problem of supply and demand and the market power lies with the professional sport teams. Most economists are agreed that this phenomenon is not an example of sport contributing substantially to economic regeneration. However, some American cities have gone beyond the professional sport team stadium game and have had a broader approach to using sport for economic regeneration. Indianapolis, Cleveland, Philadelphia, Kansas City, Baltimore and Denver are examples of cities that have adopted broader sports-orientated economic regeneration strategies and Indianapolis is perhaps the best example out of these.

Schimmel (2001) and Davidson (1999) analyse how sport has been used in Indianapolis for economic regeneration of the city. Indianapolis is a mid-western US city that in the mid 1970s was suffering from the decline of its heavy manufacturing base, in particular its car industry. Local politicians were keen to develop a new image of the city. As Schimmel indicates, the problem was not that the city had a bad image, but rather that the city had no image at all. The strategy was to target the expanding service-sector economy in an attempt to redevelop the city's downtown area by using sport as a catalyst for economic regeneration. From 1974 to 1984, a total of $1.7 billion in public and private resources was invested in inner-city construction (Schimmel, 2001) of which sporting infrastructure played a major role. The strategy included investment in facilities in professional team sports but added to this a strategy of hosting major sports events in the city.

Between 1977 and 1991, 330 sports events were hosted by Indianapolis. Davidson (1999) attempted to measure the economic contribution of sport to the city in 1991. He found that, in 1991, 18 sport organisations and 9 sport facilities in the city employed 526 people. In addition, 35 sports events held in the city in 1991 generated additional spending of $97 million. He estimated the total economic contribution of sport organisations, facilities and events in Indianapolis in 1991 to be $133 million. In addition, other studies had shown that the sport strategy aimed at economic regeneration had resulted in other non-economic benefits including increased sports participation by young people,

increased pride in the city and an enhanced image for the city resulting in more convention tourism. Although Indianapolis was an early example, the strategy of using sports events as a catalyst for urban regeneration became popular in the UK in the 1980s and 1990s.

Sport and Economic Regeneration in Cities and Regions in the UK

Several cities in the UK (Sheffield, Birmingham and Glasgow) have used sport as a lead sector in promoting urban regeneration and these three cities were awarded National City of Sport status in 1995 partly because of this. They have all invested heavily in their sports infrastructure so that each has a portfolio of major sporting facilities capable of holding major sports events.

In addition to facilities, each city has a supporting structure of expertise in event bidding and management to ensure quality bids with a high probability of success and to guarantee high-quality event management. Events are a major vehicle for attracting visitors to the city and hence contributing to urban regeneration. However, these cities are also involved with developing sport in the cities through performance and excellence programmes (such as training, squad preparation, coaching) and in community sports development, so that the local population benefits from the investment in sports infrastructure.

These and other cities have made a specific commitment to public investment in sport as a vehicle for urban regeneration. However, the quantity and distribution of returns to such public-sector investment in sport, predominantly from local government, have been largely underresearched and remain uncertain. Often such investment attracts criticism because of media attention on a specific event, such as the World Student Games in Sheffield in 1991, and there has been little research on the medium- and long-term returns on such investment.

In a report commissioned by UK Sport, *Measuring success 2: the economic impact of major sport events* (UK Sport, 2004), the Sport Industry Research Centre presented an overview of the findings from 16 economic impact studies of major sports events undertaken since 1997, many of which took place in these 3 cities and all but 3 (Spar Europa Cup, World Cup Triathlon, World Indoor Athletics) of which were carried out by the Sport Industry Research Centre. This consolidated piece of research builds on the original document published by UK Sport in 1999 (1999a), which recognised and demonstrated the potential of major sports events to achieve significant economic impacts for the cities that host them.

These 16 studies have been conducted using essentially the same methodology as published by UK Sport in 1999 entitled *Major events: the economics—a guide* (UK Sport, 1999b). This therefore provides a dataset in which the events are directly comparable and we concentrate on these comparisons. Key findings from the research are outlined in Table 1, commencing with the impact of each event.

Overall, the findings confirm that major sports events can have significant economic impacts on host communities. These impacts ranged from the £0.18 million of additional expenditure attributable to the half-day

IAAF Grand Prix Athletics staged on a Sunday in Sheffield in June 1997, to the £25.5 million attributable to the Flora London Marathon in April 2000. Moreover, other events, most notably the World Cup Triathlon, World Indoor Athletics and Test Cricket attracted additional expenditure per day in excess of £1 million. Junior events (such as European Junior Swimming and Junior Boxing) had the least significant daily impacts, mainly because they rarely attract considerable numbers of spectators. It is interesting to note that the two events generating the highest economic impacts, the London Marathon and a cricket Test Match, were domestic events that take place annually, do not need to go through a bidding process and do not require new sporting infrastructure investment.

Economic impact is not UK Sport's rationale for attracting major events to the UK, but it is a useful device by which to justify funding an event in economic terms. The evidence suggests that as a general rule it is the expenditure by visitors to an event which contributes the majority of any additional expenditure, rather than spending by the organisers of an event.

Table 1. Economic impact of 16 major sports events

Year	Event	Host city	Event days	Impact (£ million)	Impact per event day (£ million)
1997	World Badminton	Glasgow	14	2.22	0.16
1997	European Junior Boxing	Birmingham	9	0.51	0.06
1997	1st Ashes Test Cricket England v Australia	Birmingham	5	5.06	1.01
1997	IAAF Grand Prix 1 Athletics	Sheffield	1	0.18	0.18
1997	European Junior Swimming	Glasgow	4	0.26	0.06
1997	Women's British Open Golf	Sunningdale	4	2.07	0.52
1998	European Short Course Swimming	Sheffield	3	0.31	0.10
1999	European Show Jumping	Hickstead	5	2.20	0.44
1999	World Judo	Birmingham	4	1.94	0.49
1999	World Indoor Climbing	Birmingham	3	0.40	0.13
2000	Flora London Marathon	London	1	25.46	25.46
2000	Spar Europa Cup, Athletics	Gateshead	2	0.97	0.48
2001	World Amateur Boxing	Belfast	8	1.49	0.19
2001	World Half Marathon	Bristol	1	0.58	0.58
2003	World Cup Triathlon	Manchester	1	1.67	1.67
2003	World Indoor Athletics	Birmingham	3	3.16	1.05

Spectators contributed the majority of the additional expenditure at 10 of the 16 events and such events are termed 'spectator-driven'. Further analyses revealed a strong correlation between the number of spectator admissions and the absolute economic impact of an event, which suggests that the absolute number of spectators is the key driver of economic impact.

A typical competitor spends between £55 and £60 per day at an event, of which 82 per cent is spent on subsistence (accommodation, food and drink). Cricketers at the Test Match spent the most per day of all the competitors (£113), compared with athletes at the World Half Marathon who spent the least (£42). Typical daily spend of an official was £70, of which 80 per cent was attributable to expenditure on subsistence. Competitors spend relatively little on items other than subsistence, because their days are characterised by a cycle of preparation, competition and rest which leaves little time for interaction with the local economy. Similarly, officials work long hours to ensure that events run smoothly and consequently they too have little time to get out and about locally. By contrast, daily spend of a typical media representative was around £100 (and often much more for those on expenses), with 75 per cent of this attributable to spending on subsistence (usually commercial accommodation). Moreover, daily expenditure by media personnel on other items (around £25) almost doubled that spent by the typical competitor or official. Hence, not only do events benefit from the value of media coverage but also from the relatively high additional daily expenditure of media representatives.

The daily spending of spectators varies considerably across events, ranging from £86 at the European Junior Swimming (where parents spent money on behalf of and supporting their children) to less than £10 per day at the IAAF Athletics Grand Prix. Although the absolute number of spectators is the key driver of economic impact, the average spectator (at a little under £50) spends less per day than the other groups. This is because spectators are most likely to be day-visitors and least likely to make use of commercial accommodation (hotels and guest houses), as evidenced by only 59 per cent of their daily expenditure being attributable to subsistence. However, average daily expenditure of spectators is a function of the proportion staying overnight in the host area.

Much of the economic impact referred to here is actually a redistribution of money around the UK economy, which has no lasting impact on overall GDP. However, expenditure by visitors from overseas is actually 'new' money to the UK economy in the form of invisible exports as exemplified by the Flora London Marathon which revealed a net export effect approaching £1.2 million. Events that achieve this genuine inflow of funds arguably provide a better-quality impact than those associated with the recirculation of money within the UK economy. Notwithstanding this, the local organising committees of events such as the World Half Marathon or World Indoor Athletics are unlikely to worry from where any additional expenditure originates, as long as it is forthcoming. However, they may be interested in evidence suggesting that visitors from overseas stay longer and spend more than the average visitor.

The research has revealed high approval ratings from the public for continued support of events through the Lottery. Moreover, based on evidence from 10 of the 11 part-Lottery-funded events, for every £1 of Lottery support, additional expenditure in host economies amounted to £7.23. However, Lottery support rarely covers the total costs associated with hosting an event and the return on investment figure does not allow for the additional costs incurred by local organising committees. Consequently, the impact in host economies for every £1 invested at an event will be less than £7.23.

Additional benefits have been monitored at more recent events, as organisers look beyond the direct economic impact when evaluating their events following the 'balanced scorecard' approach.

The 'public profile' of the European Short Course Swimming Championships was measured by the analysis of the television coverage for the event. This monitoring of an event's television coverage has revealed some interesting and perhaps unexpected findings. The key finding is that the event achieved television audiences that were greater than some sports generally perceived as having larger audiences than swimming. Most notably, audiences for the European Short Course Swimming Championships exceeded those for some rugby union international matches as well as prestigious events in the rugby league and cricket calendars.

The European Short Course Swimming Championships achieved coverage in 18 programmes or programme segments lasting 1087 minutes which were broadcast in the UK and mainland Europe (Shibli and Gratton, 1999). A total of nearly 8 million viewers across the UK and Europe watched coverage of the event. The highest audience share was achieved in the UK (23 per cent) and the highest TVR (television rating) was achieved in Finland, where 9 per cent of the country's population watched recorded highlights of the event.

The economic impact of the spending of visitors at this event was relatively small (around £300 000). However, the public profile achieved by the television coverage was worth substantially more than this to the host city Sheffield, the event itself (owned by the international governing body LEN) and the event sponsor (Adidas).

The analysis of these events shows the wide variety in economic impact generated by different events and how, for some events, other benefits can be greater than economic impact. Some of the events generate relatively small economic impacts. Just because the event is a world or European championship does not guarantee that it will be important in economic terms. The difficulty for cities trying to follow an event strategy for regeneration purposes is that it is difficult to forecast the economic impact of any event prior to staging it. However, cities such as Sheffield, Birmingham and Glasgow that now have a history of hosting a wide range of events do acquire the experience of being able to judge those events which generate the most significant benefits.

Case Study: Commonwealth Games Manchester 2002

The Commonwealth Games held in Manchester in 2002 involved an investment of £200 million in sporting venues in the city and a further £470 million investment in transport and other infrastructure. This is by far the largest investment related to the hosting of a specific sports event ever to be undertaken in Britain. It was also the first time in Britain that planning for the hosting of a major sports event was integrated with the strategic framework for the regeneration of the city, in particular east Manchester.

In 1999, three years before the Games were held, the Commonwealth Games Opportunities and Legacy Partnership Board was established to manage the legacy of the Games. Legacy activities were funded under the 2002 North West Economic and Social Single Regeneration Board Programme which operated from 1999 to 2004. This was the first time in Britain an ambitious legacy programme was designed around a major sports event. The objective was to ensure that the benefits of hosting the event would not disappear once the event was over but that rather there would be a long-term permanent boost to the local economy of east Manchester.

Despite the long-term planning for the Games and the legacy there was one major omission: no economic impact study was carried out during the Games in 2002 and so no primary data are available on the immediate economic benefit of the Games. Cambridge Policy Consultants produced a pre-event estimate of the economic impact in April 2002 and then revised it in November 2003 (Cambridge Policy Consultants, 2003) using secondary evidence available from the Games period. They estimated that the Games generated 4900 FTE additional jobs in Manchester. However, without any visitor survey data available for the Games themselves, there must be serious doubts as to the validity of such an estimate.

A further study of the benefits of the Games was carried out for the North West Development Agency in 2004 by Faber Maunsell, in association with Vision Consulting and Roger Tym and Partners (Faber Maunsell, 2004). The study used secondary sources and interviews with key stakeholders.

As part of the study, they measured employment change in east Manchester between 1999 and 2002 as revealed by the Annual Business Inquiry (ABI) data. This showed a 1450 increase in jobs (including both part-time and full-time jobs) or a 4 per cent increase over the 1999 level. However, these are annual data and therefore it is difficult to isolate how much of this increase was due to the Games. The distribution of the increase in construction (23 per cent increase), distribution, hotels and restaurants (14 per cent increase) and other services (24 per cent increase) is consistent with the Games having been the main generator of the increase in jobs. Also, out of the 210 new jobs in 'other services', 200 were in the 'recreational, cultural, and sporting' category, suggesting again a significant Games effect. However, 1450 new jobs, which included part-time jobs, is considerably different from the 4900 FTE jobs estimated by Cambridge Policy Consultants, although this figure relates to the effect on the whole of Manchester not just east Manchester.

The net additional value of capital investment in the Games was estimated by Faber Maunsell at £670 million, of which £201 million was for the sporting venues and £125 million was for transport infrastructure. Other major investment included an Asda-Walmart superstore occupying 180 000 square feet and employing 760 FTE staff.

Since no visitor survey was carried out during the Games, actual tourism indicators were difficult to obtain. Using annual tourism data from the UK Tourism Survey (UKTS) and the International Passenger Survey (IPS), Faber Maunsell (2004) indicate a 7.4 per cent increase of overseas-resident visitors to Greater Manchester in 2002

compared with 2000. However, there was a 6.4 per cent decrease in UK-resident visitors to Greater Manchester over the same period and a 2.2 per cent decrease in the number of nights overseas residents spent in Greater Manchester. Overall, though, there was a 21 per cent increase in UK-resident expenditure and a 29 per cent increase in overseas-resident expenditure in Greater Manchester in 2002 compared with 2000. Again, because these are annual figures, it is impossible to isolate the influence of the Games, although it is reasonable to conclude that they were the most significant factor.

The Faber Maunsell study does not give a detailed media analysis of the Games, indicating only that the opening and closing ceremonies had an 'estimated' world-wide audience of 1 billion. The Commonwealth Games is an unusual event in that it does get television coverage across most continents but is not a global event in the same way as the Olympics and the football World Cup are. There are key markets where there will be no coverage at all. These include the US, the whole of the rest of Europe outside the British Isles, Japan and China. The event, therefore, is limited in terms of its potential effect on the image and profile of the host city.

Some indication of the public profile benefits of the Games is indicated by Manchester moving up the European Cities Monitor from 19th in 2002 to 13th in 2003. The Monitor is a measure of the best European cities in which to locate a business compiled by Cushman and Wakefield Healey and Baker. It is constructed from the views of Europe's 500 leading businesses on the top business locations in Europe and is used to indicate aspects affecting business location decisions. For Manchester, it is an indicator of an improvement in the city's image from a business perspective and an indicator of greater potential for inward investment.

Despite the lack of hard evidence on the economic impact of the Commonwealth Games on Manchester in 2002, there is enough evidence to indicate that east Manchester has benefited considerably. Manchester City now use the City of Manchester stadium as their home ground and other sporting venues in east Manchester have become the English Institute of Sport and are used for the training of élite athletes. Since much of the funding for the new investment for the facilities came from the National Lottery or central government, this is a clear economic boost for the area. We will have to wait and see whether the legacy benefits are as great as were hoped for, but the indications are promising.

Longer-term Benefits of Hosting Major Sport Events

Although it is too early to assess the urban regeneration legacy benefits of Manchester 2002, it should be possible to assess the long-term benefits of events held 10 or 20 years ago. Unfortunately, there are few research studies that attempt to measure systematically such long-term benefits. Spilling (1998) found that he could identify no long-term economic benefits for Lillehammer from hosting the Winter Olympics in 1994. He concluded that

> If the main argument for hosting a mega-event like the Winter Olympics is the long-term economic impacts it will gene-rate, the Lillehammer experience quite clearly points to the conclusion that it is a waste of money (Spilling, 1998, p. 121).

Spilling seems to question whether there can be any long-term effect for a area the size of Lillehammer, a city of 24 000 inhabitants situated 180 kilometres north of Oslo. The two Winter Olympics prior to the Lillehammer Games, in Calgary in 1988 and in Albertville in 1992, had been in larger regions and there was more evidence of a continuing benefit several years after the Games. In the case of Albertville, this was partly due to massive transport infrastructure investment which made access to the region by car substantially easier—although at a severe cost to the alpine environment. It is certainly the case that there is little evidence to support the argument that the Winter Olympics leave a substantial long-term benefit.

There is some evidence, however, that the Summer Olympics do generate a legacy benefit. One example that is often quoted to support the argument that there are long-term benefits of hosting major sports events is the case of the Barcelona Olympics in 1992. Sanahuja (2002) analysed the benefits to Barcelona in 2002, 10 years after hosting the games. Table 2 shows almost a 100 per cent increase in hotel capacity, number of tourists and number of overnight stays in 2001 compared with the pre-Games position in 1990. Average room occupancy had also increased from 71 per cent to 84 per cent. In addition, the average length of stay had increased from 2.84 days to 3.17 days. In 1990, the majority (51 per cent) of tourists to Barcelona were from the rest of Spain, with 32 per cent from the rest of Europe and the remainder (17 per cent) from outside Europe. By 2001, the absolute number of Spanish tourists had actually risen by 150 000 but, given the near-doubling in the number of tourists overall, this higher total only accounted for 31 per cent of the total number of tourists. The proportion of tourists from the rest of Europe went up from 32 per cent to 40 per cent (representing an absolute increase of around 800 000) and from the rest of the world from 17 per cent

Table 2. Legacy benefits of the Barcelona Olympic Games

	1990	2001
Hotel capacity (beds)	18 567	34 303
Number of tourists	1 732 902	3 378 636
Number overnights	3 795 522	7 969 496
Average room occupancy (percentage)	71	84
Average stay	2.84	3.17
Tourist by origin (percentages)		
Spain	51.2	31.3
Europe	32	39.5
Others (US, Japan, Latin America)	16.8	29.2

Source: Turisme de Barcelona (BarcelonaTourist Board) and Sanahuja (2002).

to 29 per cent (representing an absolute increase of around 600 000).

Overall infrastructure investment prior to the Games was $7.5 billion compared with a budget of around $1.5 billion for the Olympic Committee to stage the games. The Olympics in Barcelona were the most expensive ever staged. However, Barcelona's use of the Games as a city marketing factor is generally regarded as a huge success. This is evidenced by Barcelona's rise in ranking in the European Cities Monitor from 11th in 1990 to 6th in 2002.

Given the scarcity of evidence on the long-term urban regeneration benefits of hosting sporting events, the Department of Culture, Media and Sport/Strategy Unit in their review of sport strategy in England were sceptical over the existence of such benefits

> Our conclusion is that the economic justifications for any future bids for mega-events must be rigorously assessed. If regeneration is intended as an explicit pay-off from hosting a mega-event, then it must underpin the whole planning process to ensure that maximum benefit for the investment is achieved (Department of Culture, Media and Sport/Strategy Unit, 2002, p. 68).

It is interesting, therefore, that very soon after this review was published in December 2002, the government decided to back the bid for London to stage the 2012 Olympics, which tends to support Roche's (1994) argument that in the end such decisions are political rather than part of a rational planning process.

Regional Development Agencies (RDAs) and Sport-related Economic Regeneration

Although the emphasis in this paper has been on the role of sport in the economic regeneration of cities, over the recent past the creation of Regional Development Agencies (RDAs) in the UK in the late 1990s with specific responsibility for economic regeneration has led to sport being looked on as a catalyst for regional economic development. This has led to an emphasis away from just the

staging of events to a broader-based sport industry strategy for regeneration purposes.

The one area of sport to have the most recognition for its economic potential by RDAs is the motorsport industry. The National Survey of Motorsport Engineering and Services carried out by the Motorsport Industry Association (MIA) (2001) revealed that 4200 UK businesses are involved with a total annual turnover of £4.6 billion and total full-time employment of 38 500. The survey revealed that the industry was a young industry with three-quarters of the engineering firms involved having been set up in the past 20 years. Most are independently owned and have an average of 23 employees.

Motorsport Valley, as the motorsport cluster has been called, spans 4 RDA areas—East and West Midlands, East of England and the South East—these areas accounting for 77 per cent of all UK motorsport businesses. The East Midlands Development Agency leads on the development of the motorsport industry on behalf of all the RDAs in Motorsport Valley.

In July 2003, the Department of Trade and Industry announced the launch of a new £16 million government fund to accelerate motorsports into first place by strengthening the industry and boosting all areas from engineering to Formula 1. The cash will help put into place recommendations made by the Motorsport Competitiveness Panel which includes an RDA-led Government Motorsport Unit, which would act as a one-stop-shop in monitoring and giving advice on funding to motorsport companies. The recommendations comprise joint industry, sport and government initiatives to improve motorsport skills, education and training services; encourage increased motorsport technology and business development; and also ways to encourage wider participation in the sport by making it more accessible and socially inclusive.

In November 2003, the Motorsport Development Board (MDB) replaced the strategically focused Motorsport Competitiveness Panel. The role of the MDB over the next five years will be to implement recommendations set out in the Panel's final report to sustain and develop the UK motorsport cluster.

The only RDA specifically to encompass the whole of the sport indusry in its cluster development strategy is the North West Development Agency. This was partly triggered by the success of the 2002 Commonwealth Games but also by an earlier study in 1999 which identified the North West region as having a potential competitive advantage in the sport business sector. A sport cluster study has been commissioned by the NWDA in addition to a major events strategy.

The North East Development Agency, One NorthEast, although not at the moment identifying sport as a specific business cluster, has acknowledged the potential of sport's contribution to the economic development agenda by appointing a senior executive for sport and economic development. Following this, a study of the economic potential of sport in the North East Region was commissioned early in 2003. Since that, a further study has been commissioned into the economic importance of golf tourism in the region.

Finally, the South West Development Agency, in partnership with other funding agencies, commissioned a study of the economic impact of sport in Cornwall and the Isles of Scilly in 2004. This study (Cornwall Enterprise, 2004) highlighted the importance of the surfing industry in Cornwall with an annual turnover of £64 million and with 1036 full-time jobs and 571 part-time jobs.

These examples show that RDAs are only now starting to investigate the potential of sport for economic regeneration. The example of the motorsport sector is at a more advanced stage of development with the commitment of £16 million of central government funding for this sector.

Conclusions

Sport has the potential to generate substantial economic and social returns to local and regional government investment in the sports industry. The research focus over the past decade, however, has been the national economic importance of sport. Although some evidence is available on the economic benefits of sports events, and sports tourism, many of the economic benefits to the local community have been poorly researched. Most of the serious gaps in knowledge over the broader economic benefits of sport can best be filled at the local level. Such research would allow more rational investment appraisal in new investments in sports infrastructure and sports programmes by local government.

It is clear from the discussion in this paper, however, that both in North America and Europe the strategic thinking relating economic regeneration and sport has been dominated by the view that sport can only contribute to economic activity by attracting sports tourists, either spectators or participants, to the city or region. Such strategies have also been relatively easy to 'sell' to taxpayers in the local economy since the economic argument has been reinforced by the additional generation of social and environmental benefits that such a sport-led economic regeneration can bring to local residents and taxpayers.

In North America, there is increasing questioning over the investment of public money into professional team sports that generate huge profits to their owners and athletes. In Europe, however, economic impact studies over the recent past have shown there to be a small number of major sports events (including the Olympics, the World and European Championships in football) that generate an unequivocal economic benefit to host cities. There is another group of events (such as Wimbledon, the FA Cup Final, Six Nations Rugby Internationals) that also generate significant economic benefits but are not normally 'on the market' for competing cities to bid for (i.e. they always take place in the same venues each year). There are a large number of other events (National, European and World Championships across all sports) that have the potential to generate significant economic impact. The evidence provided in this paper has shown the wide diversity in economic impacts generated from such events, but also that a sport strategy based around events can deliver significant benefits to cities.

Whether such benefits justify the expenditure involved is, however, a difficult question to answer. When the money for sporting infrastructure investment is provided by local taxpayers, as it was for the World Student Games in Sheffield, the question arises of whether other projects might have provided better returns to the local community. When the money for investment comes primarily from outside the local community, as it did for the Commonwealth Games in Manchester, then it is an unequivocal benefit to the local community in economic terms, but may not be the best use of the funds from a national perspective. At this point in time, we simply do not have adequate evidence to make judgements of this type. The evidence that we do have relates to the immediate economic impact during and immediately after the event has been held. There is a need for research to concentrate on the longer-term urban regeneration benefits that sport has the potential to deliver.

References

BAADE, R. A. (1996) Professional sports as catalysts for economic development, *Journal of Urban Affairs*, 18(1), pp. 1–17.

BAADE, R. A. (2003) Evaluating subsidies for professional sports in the United States and Europe: a public sector primer, *Oxford Review of Economic Policy*, 19(4), pp. 585–597.

BIANCHINI, F. and SCHWENGEL, H. (1991) Re-imagining the city, in: J. COMER and S. HARVEY (Eds) *Enterprise and Heritage: Cross-currents of National Culture*, pp. 214–234. London: Routledge.

BRAMWELL, B. (1995) *Event tourism in Sheffield: a sustainable approach to urban development?* Unpublished paper, Centre for Tourism, Sheffield Hallam University.

BURNS, J. P. A., HATCH, J. H. and MULES, F. J. (Eds) (1986) *The Adelaide Grand Prix: the impact of a special event*. The Centre for South Australian Economic Studies, Adelaide.

CAMBRIDGE POLICY CONSULTANTS (2003) *The Commonwealth Games 2002: a cost and benefit analysis: executive update*.

COATES, D. and HUMPHREYS, B. (1999) The growth of sports franchises, stadiums and arenas, *Journal of Policy Analysis*, 18(4), pp. 601–624.

CROMPTON, J. L. (1995) Economic impact analysis of sports facilities and events: eleven sources of misapplication, *Journal of Sport Management*, 9(1), pp. 14–35.

CROMPTON, J. L. (2001) Public subsidies to professional team sport facilities in the USA, in: C. GRATTON and I. P. HENRY (Eds) *Sport in the City: The Role of Sport in Economic and Social Regeneration*, pp. 15–34. London: Routledge.

CROMPTON, J. L. (2004) Beyond economic impact: an alternative rationale for the public subsidy of major league sports facilities, *Journal of Sport Management*, 18, pp. 40–58.

DAVIDSON, L. (1999) Choice of a proper methodology to measure quantitative and qualitative effects of the impact of sport, in: C. JEANREAUD (Ed.) *The Economic Impact of Sport Events*, pp. 9–28. Centre International d'Etude du Sport (CIES), Neuchatel.

DEPARTMENT OF CULTURE, MEDIA AND SPORT/ STRATEGY UNIT (2002) *Game plan: a strategy for delivering Government's sport and physical activity objectives*. London.

DEPARTMENT OF THE ENVIRONMENT (1975) *Sport and Recreation*. Cmnd 6200/. London: HMSO.

FABER MAUNSELL (2004) *Commonwealth Games Benefit Study: Final Report*. Manchester: North West Devlopment Agency.

GIBSON, H. J. (1998) Sport tourism: a critical analysis of research, *Sport Management Review*, 1, pp. 45–76.

GRATTON, C. and TAYLOR, P. (1991) *Government and the Economics of Sport*. Harlow: Longman.

KASIMATI, E. (2003) Economic aspects and the Summer Olympics: a review of related research, *International Journal of Tourism Research*, 5, pp. 433–444.

LOFTMAN, P. and SPIROU, C. S. (1996) *Sports stadiums and urban regeneration: the British and United States experience*. Paper to the Conference, Tourism and Culture: Towards the 21st Century, Durham, September.

MOTORSPORT INDUSTRY ASSOCIATION (2001) *The National Survey of Motorsport Engineering and Services*. Stoneleigh Park, Warwickshire.

MULES, T. and FAULKNER, B. (1996) An economic perspective on major events, *Tourism Economics*, 12(2), pp. 107–117.

NOLL, R. and ZIMBALIST, A. (Eds) (1997) *Sports, Jobs & Taxes*. Washington, DC: The Brookings Institution.

QUIRK, J. and FORT, R. (1999) *Hard Ball: The Abuse of Power in Pro Team Sports*. Princeton, NJ: Princeton University Press.

RITCHIE, J. R. B. (1984) Assessing the impact of hallmark event: conceptual and research issues, *Journal of Travel Research*, 23(1), pp. 2–11.

RITCHIE, J. R. B. and AITKEN, C. E. (1984) Assessing the impacts of the 1988 Olympic Winter Games: the research program and initial results, *Journal of Travel Research*, 22(3), pp. 17–25.

RITCHIE, J. R. B. and AITKEN, C. E. (1985) OLYMPULSE II—evolving resident attitudes towards the 1988 Olympics, *Journal of Travel Research*, 23(Winter), pp. 28–33.

RITCHIE, J. R. B. and LYONS, M. M. (1987) OLYMPULSE III/IV: a mid-term report on resident attitudes concerning the 1988 Olympic Winter Games, *Journal of Travel Research*, 26(Summer), pp. 18–26.

RITCHIE, J. R. B. and LYONS, M. M. (1990) OLYMPULSE VI: a post-event assessment of resident reaction to the XV Olympic Winter Games, *Journal of Travel Research*, 28(3), pp. 14–23.

RITCHIE, J. R. B. and SMITH, B. H. (1991) The impact of a mega event on host region awareness: a longitudinal study, *Journal of Travel Research*, 30(1), pp. 3–10.

ROCHE, M. (1992) Mega-event planning and citizenship: problems of rationality and democracy in Sheffield's Universidade 1991, *Vrijetijd en Samenleving*, 10(4), pp. 47–67.

ROCHE, M. (1994) Mega-events and urban policy, *Annals of Tourism Research*, 21(1), pp. 1–19.

SACK, A. and JOHNSON, A. (1996) Politics, economic development and the Volvo International Tennis Tournament, *Journal of Sport Management*, 10, pp. 1–14.

SANAHUJA, R. (2002) *Olympic City—the city strategy 10 years after the Olympic Games in 1992*. Paper delivered to the *Conference, Sports Events and Economic Impact*, Copenhagen, April 2002.

SCHIMMEL, K. S. (2001) Sport matters: urban regime theory and urban regeneration in the late capitalist era, in: C. GRATTON and I. P. HENRY (Eds) *Sport in the City: The Role of Sport in Economic and Social Regeneration*, pp. 259–277. London: Routledge.

SHIBLI, S. (2001) Using an understanding of the behaviour patterns of key participant groups to predict the economic impact of major sports events, in: *Proceedings for the 9th EASM Congress*, pp. 294–298. Vitoria-Gasteiz, Spain.

SHIBLI, S. and GRATTON, C. (1999) Assessing the public profile of major sports events: a case study of the European Short Course Swimming Championships, *Sports Marketing & Sponsorship*, 1, pp. 278–295.

SPILLING, O. A. (1998) Beyond intermezzo? On the long-term industrial impacts of mega-events: the impact of Lillehammer 1994, *Festival Management & Event Tourism*, 5, pp. 101–122.

UK SPORT (1999a) *Major events: the economics—measuring success*. London.

UK SPORT (1999b) *Major events: the economics—a guide*. London.

UK SPORT (2004) *Measuring success 2: the economic impact of major sports events*. London.

Just Art for a Just City: Public Art and Social Inclusion in Urban Regeneration

Joanne Sharp, Venda Pollock and Ronan Paddison

Introduction

As part of the celebration of Scottish devolution, implemented in 1999, but also as a sculpture intended to be a part of the restructuring of the 'new Glasgow', a statue to Donald Dewar was erected at a prominent location in the city centre. Generally acclaimed as the 'father' of Scottish devolution, but also an MP of long standing in the city, a statue to him seemed a fitting celebration of his achievements. Moreover, its emplacement at the head of a newly pedestrianised area and immediately outside the new Concert Hall, itself a product of the city's status as European City of Culture in 1990, seemed an appropriate gesture to both the city and Scotland. Yet, repeatedly, the statue has been vandalised, to the point that the city council considered relocating it, or at least, through raising the height of the plinth on which it stands, making its vandalism more difficult.

As it is, the statue remains at its original site, although as a result of vandalism the subject often lacks his spectacles and is periodically embellished with *graffiti*.

The story the Donald Dewar statue tells is one repeated elsewhere, that public art can be read in different ways and that its uses to beautify the city or celebrate its reimagineering do not necessarily enjoy universal consensus. In this respect, public art is no different from art in general where matters of taste and preference become paramount. For public art, these issues become magnified precisely because of its visibility and hence its 'inescapability', although reactions to it can vary from the highly vocal and oppositional to the unaffected. Time can help to mellow public opinion to artworks so they become part of not just the taken-for-granted but also of the accepted landscape of the city. A few years before the unveiling of the statue to

Donald Dewar, public opposition had been vocal to the suggestion made by Glasgow City Council to relocate statues from the city's principal square. Yet, ironically, the statues commemorate largely forgotten political and military figures of the 19th century and are symptomatic of the 'imposed monumentalism' of the Victorian city. What Lefebvre (1991, p. 143) warned of as the ability of "monumental buildings to mask the will to power and the arbitrariness of power beneath signs and surfaces which claim to express collective will and collective thought", is given added weight through the impress of time and habituation.

Such contradictions underline the different readings public art attracts, but they also suggest how its meaning, for the self and more specifically the self as citizen, can be read as more or less inclusive. Where exclusion reflects authoritarian imposition, it is in the colonial city that the alienating effects generated by public art, particularly that celebrating imperial control, foster political reaction and the will to decommemorate alien rule. Thus, following Irish independence, the monumental symbols to British rule in Dublin were successively removed, sometimes by the state at others clandestinely by nationalist groups. Yet as Whelan (2003) shows in the case of the most obvious of the icons, certainly the most prominent in the urban landscape—Nelson's Pillar modelled on its London counterpart—opinions as to its fate were divided. Ultimately it was to be decided by the bomb. Yet in spite of the overt political symbolism of the pillar, its familiarity and acceptance as part of the everyday use of the city, as well as appreciation of its aesthetic qualities, meant that its removal was not uncontentious. As Whelan has argued

> With the passage of time it became a popular meeting-place and viewing point ... and a symbol of the city centre that transcended any political connotations (Whelan, 2003, p. 206, emphasis added).

Such decommemoration is commonplace in cities emergent from periods of authoritarian rule—although not necessarily uncontested

as can be the remonumentalisation of (urban) space (Czaplicka and Ruble, 2003).

Where the development of public art as part of the repertoire of the gentrification of the contemporary city lacks the overt political symbolism of monumentalism (Levinson, 1998), this is not to suggest that its use is politically neutral. Deutsche (1996) argues forcefully how the promotion of public art and architecture appears to neutralise politically its use within the city yet masks its political outcomes, particularly on those excluded from the new image created. Contemporary trends in public art in the city have tended to eschew monumentalism as it was expressed in the 19th century with its thinly disguised appeal to élite interests. Further, much as public art and architecture in Rome or Florence in the Reformation had been fashioned to celebrate the city and in the 19th century became part of the process of forging the City Beautiful, so its present use, in part at least, can be seen as part of the ongoing goal of beautifying the city. Yet the (re)aestheticisation of cities is not an apolitical exercise; the Hausmannisation of European cities in the 19th century and its 'imposed' nature and socially divisive outcomes have their parallels in the contemporary restructuring of the city under what Harvey (2000) has described as 'neoliberalised urban authoritarianism'. Much as historical analogy risks glossing over contextual differences, what Gunn (2000) has sought to demonstrate as the dominance of bourgeois values on the landscapes of the cities of northern England in the 19th century has come to be repeated albeit in a different guise in the bourgeois revanchism underpinning contemporary urban gentrification (Smith, 1996).

What the experience of urban regeneration continues to repeat is that the uses to which culture has been employed as part of the process of revival can be socially divisive leading to what Mitchell (2000) has described as 'culture wars'. As has been widely remarked upon (for example, see Bianchini, 1999; Boyle and Hughes, 1991), cultural planning immediately raises the question of 'culture for whom?' in which imposition and the favouring of particular interests are likely to engender reaction and resistance. To its practitioners, there may be a degree of inevitability here, particularly where the reimagineering of cities has become so focused around the winning of mega-events; the focused nature of such events may be incompatible with the ability to address the diverse set of preferences represented in the city. Yet where 'culture wars' arise, their development reflects the wider problem (and challenge) apparent in contemporary urban restructuring in which the battleground of city politics comprises two 'sides' of urban entrepreneurialism caught up in the avowed objective of making the city more competitive and the increasing social inequalities that have become so much the hallmark of such cities.

In this article, we want to show how cultural policy, and in particular public art, intersects with the processes of urban restructuring and how it is a contributor, but also antidote, to the conflict that typically surrounds the restructuring of urban space. Even restricting our attention to the field of public art, it is apparent that these intersections are complex and contextually dependent. Our intention is to 'cut into' the picture through asking two sets of questions linked to the overarching purpose of investigating how public art can be inclusionary/exclusionary in the methods through which it has been practised as part of the wider project of urban regeneration.

—In the deployment of public art, what conditions contribute to or hinder democratically inclusive practices? How is local participation able to counter top–down practices? How do design professionals, architects and artists, seek to develop inclusive practices? Is inclusion seen as an end in itself or a means to an end, and by whom?
—In what ways have the claims made for the use of public art within urban regeneration been inclusive? Under what conditions does inclusion contribute to a sense of democratic ownership over the inscription of urban spaces? How does this vary between different types of public art and

architecture, and in different types of urban space?

Fundamentally, our concern in this paper is to offer critical insight into how public art and architecture contribute or otherwise to the social cohesion of the city. Key to the creation of social cohesion is the belief that public art, or the processes through which it is produced, is able to create a sense of inclusion. By this token, public art should be able to generate a sense of ownership forging the connection between citizens, city spaces and their meaning as places through which subjectivity is constructed. Initially, we outline the rationale for identifying public art and the critiques that are used to counter the rhetoric underpinning its adoption. Here, we are concentrating on the visual so excluding those other elements of the arts, notably the performative, that other writers have sought to include within its definition (Deutsche, 1996; Lippard, 1997; Miles, 1997). The main discussion is divided into two sections looking at different types of in(ex)clusive practice through different case studies. Initially, we look at examples in which public art intervention has been sought inclusively. Subsequently, attention focuses more on examples in which cultural domination has provoked resistance. Throughout, it is argued that in the deployment of public art it is the processes through which it becomes installed into the urban fabric that are critical to inclusion. However, practice also emphasises differences between the motivations underpinning the use of public art, the scale of intervention and the perceived importance of it to the reaestheticisation of urban spaces.

Why Public Art?

Public art is not simply art placed outside. Many would argue that traditional gallery spaces are public in their openness to interested viewers, while, conversely, others would insist that the privatisation of public space has meant that art placed in public space is not necessarily for all. Thus, public art is art which has as its goal a desire to engage with its audiences and to create spaces—whether material, virtual or imagined—within which people can identify themselves, perhaps by creating a renewed reflection on community, on the uses of public spaces or on our behaviour within them. Public art, then, does not have only to be expressed visually. It can be expressed in terms of soundscapes, media (non-)places such as the Internet, on television, as well as in material spaces of inhabited landscapes. However, given the focus here on the links between arts and urban regeneration, we have chosen to concentrate on the visual. The core examples relate to the urban environment, yet consideration is also made of some that may lie somewhat outwith the urban realm but are integrally related to ideas of ownership, identity and the creation of space.

In the UK, as in many other contemporary Western countries, public art appears to have an increasingly prominent role in urban design. In 1993, around 40 per cent of local authorities in the UK had adopted a public art policy of some sorts (Miles, 1997, p. 96) and since then progressively more cities, like Newcastle and Gateshead, have been using public art as a keystone in their regeneration schemes. Hall and Robertson (2001, p. 7) cite the general aim of adding an "'aura' of quality", listing the Policy Studies Institute's summary of the contribution that public art can make to a number of contemporary urban issues: contributing to local distinctiveness; attracting investment; boosting cultural tourism; enhancing land values; creating employment; increasing use of urban spaces; and, reducing vandalism. For its advocates, there is an overall sense of the significant role that public art can play in culture-led urban regeneration, in the economic realm, but also in terms of culture and community.

It is perhaps the perceived potential of public art to work on multiple levels and its adaptability that gives it such cultural viability. Public art not only contributes to the visual attractiveness of the city and has the ability to aestheticise urban spaces, but also, through public art, authorities can signal their willingness to deal with social and environmental problems. For many

authorities, inclusive, community-based projects appeal because they are generally low-cost and yet are perceived to be able to yield benefits beyond the aesthetic that correlate with social policy objectives. However, the way in which such projects are inscribed into regeneration policy has implications, as Phillips (1988) suggests, for the potential of the artwork. The arts represent

the more intangible phenomenon whereby cultural resources are mobilised by urban managers in an attempt to engineer consensus amongst the residents of their localities, a sense that beyond the daily difficulties of urban life which many of them might experience the city is basically doing 'alright' by its citizens (Philo and Kearns, 1993, p. ix).

Phillips (1988) and Deutsche (1996) have been quick to point out that the notion of the public should not be regarded as a neat, always consensual affair. Many arguments are based on essentialist claims to nature, identity, place and community. They thus fail to acknowledge the contested, fragmented and mutable nature of these concepts. Deutsche (1996, p. 270) worries that those who see public art as leading to the enhancement of community miss the point in that they "presume that the task of democracy is to settle, rather than sustain, conflict". Public space and the controversies surrounding public art can only reflect their constituent communities. Hall and Robertson (2001, p. 19) argue that the role of public art should be to encourage the sound of contradictory voices—voices that represent the diversity of people using the space—rather than aspire "to myths of harmony based around essentialist concepts". Phillips further points to the bureaucracy that so much public art now has to negotiate given the intended goals of inclusion—from the different committees that must examine and accept proposals to considerations of health and safety—that any critical edge is lost and the resultant work must be bland, engaging everyone but offending no-one. She says

Isn't it ironic that an enterprise aimed even at the least, at enlivening public life is now running on gears designed to evade controversy (Phillips, 1988, p. 95).

Approaching Inclusion

One of the more pressing issues characterising contemporary cities—certainly one which preoccupies much academic and policy debate—is how to achieve greater social inclusion in cities which, locked into the task of enhancing their competitive position in an increasingly globalised economy, are characterised by deepening socioeconomic inequalities and increasing segregation. The apparent unambiguity of the issue belies the complications to which it gives rise. How is inclusion to be defined? How is it to be sought? What are the presumed linkages between social inclusion and urban economic competitiveness? Such fundamental issues problematise not only how inclusion should be formulated within urban policy but also its purpose and benefits—questions that recur within the use of public art as part of the process of regenerating the city and its neighbourhoods.

As it has been suggested, even the linkages between social inclusion and urban economic competitiveness are disputed. To some (for example, Marcuse and van Kempen, 2002), social inclusion, or rather its antonym social exclusion, erodes the ability of the city to be competitive. Cities characterised by deep socioeconomic inequalities would be less attractive to investment capital undermining their ability to maintain their competitiveness. Further, as an argument to which New Labour's urban policy has given explicit support (Imrie and Raco, 2003), social inclusion was not only important to the attainment of economic competitiveness but through the Third Way the achievement of both was possible (Giddens, 1998). Not only would the opportunities to participate be enhanced in a prosperous (urban) economy, but also the benefits of growth would trickle down the social hierarchy. Inclusion, then, becomes a necessary part of a virtuous cycle

of urban growth, an argument which was to be tested in different ways in the recent research project *Cities: Competitiveness and Cohesion Research Programme* funded by the Economic and Social Research Council in the UK. The evidence of the programme was far from supportive of the linkage. Indeed, as the authors to the report summarising the programme suggested, from the evidence of Britain's 'successful cities', such as London, Leeds, Bristol or Edinburgh,

> it is clear that competitive success is far from incompatible with persistent concentrations of unemployment and social deprivation and high levels of social and economic inequality (Boddy and Parkinson, 2004, p. 428).

What the research identified was what has been empirically demonstrated elsewhere, that urban economic restructuring is often accompanied by deepening socioeconomic inequalities (Sassen, 2001; Madanipour *et al.*, 1998). Even if, as Moulaert *et al.* (2003) suggest, the impress of such inequalities varies according to the type of welfare regime and the regulatory frameworks through which urban policy is mediated, such differences do not negate the uncertainty surrounding the linkages between social inclusion and economic change.

These uncertainties become replicated in debates on the role of public art in urban regeneration. Indeed in the case of public art, doubts surround not only the contribution it might make to urban economic growth, but also to that of social inclusion. Two interrelated factors help to explain the problems: first, the contribution of public art is often deliberately symbolic; and, secondly, following from this, there are methodological problems in evaluating its impacts. Typically, the outcomes of social inclusion as part of urban policy become expressed in material terms. Most urban policy is aimed at reducing material inequalities—for example, through neighbourhood regeneration, the rehabilitation of sub-standard housing stock, training programmes aimed at reinserting the unemployed within the labour market and through the quest to improve the delivery of public services in disadvantaged neighbourhoods. As the consequences of public art are perceived to be symbolic rather than material, this tends to increase the conflict surrounding its use, which in turn is amplified by the difficulties in measuring the benefits which are claimed for it (Selwood, 1995). There are exceptions, notably what Plaza (2000) terms the 'Guggenheim effect': the contribution of iconic architecture to the generating of urban tourism and more generally to the regeneration of the city. Most public art, however, is more modest in its intervention and scale and its economic contribution is often marginal and typically indirect. The indeterminacy of its economic contribution places added attention on its imputed non-material benefits. The intangibility and contested nature of these benefits, how they contribute to building social inclusion, shifts the emphasis from outcomes towards the processes through which public art is produced and how these can foster a sense of inclusion. In other words, it is by focusing attention on the democratic processes through which public art is produced and the extent to which these are inclusive that we can begin to appreciate the role of public art in urban regeneration.

Recent debates amongst post-modern political philosophers provide pointers to how democratic processes can (and should) be more inclusive. Young (2000) provides a succinct definition of inclusion as

a democratic decision (being) normatively legitimate only if all those affected by it are included in the process of discussion and decision-making (Young, 2000, p. 22).

Her emphasis is on the processes through which collective decisions are made. Critical here is that the processes in which discussion and deliberation amongst the multiple groups and communities comprising the city afford equal status to each and that debate becomes the means of exposing and being more responsive to difference. Others, notably Fraser (1995, 1997) but also Phillips (2004), have taken issue with the emphasis given by

Young to the politics of difference and the extent to which the overaccentuation of what Fraser defines as cultural injustice has diverted attention from socioeconomic redistribution. As important as such debates are (and they become mirrored elsewhere within the use of culture as a means of urban regeneration), they should not be allowed to overshadow the common ground that exists—the commitment to social justice and the contribution to its attainment through the need for mutual recognition between groups with different preferences, the acceptance of difference and the role deliberative processes of political interaction can play.

Fraser (1995, p. 71) suggests that the processes through which cultural (or symbolic) injustices tend to arise are fundamentally "rooted in social patterns of representation, interpretation and communication". Developing this, Fraser identifies three interrelated practices commonly associated with cultural injustice. These include *non-recognition* which renders groups invisible "via the authoritative, representational and interpretative practices of one's own culture" and *disrespect*, the routine malignment "in stereotypical public cultural representations and/or in everyday life interactions". Both are fundamental to the overarching injustice of *cultural injustice*, of "being subjected to patterns of interpretation and communication that are associated with another culture and/or hostile to one's own" (Fraser, 1995, p. 71). In short, in a democratic society, equal status must be given to individuals and groups, an idea that should saturate its practice as well as being apparent in its cultural outcomes.

While these dimensions overlap, the paper uses each in turn to discuss how public art has been used to foster social inclusion in the city. Collectively, they provide pointers to what, in public art terms, would define an inclusive city, as one giving expression to the multiple and shifting identities of different groups, as indicative of presence rather than absence, and of avoiding the cultural domination of particular élites or interests. Such a mapping represents an ideal. The reality of cities, their social diversity and fluidity, and the power relations underpinning the uses of public art challenge the ability of meeting such ideals.

Our approach is empirical, drawing on specific examples of how public art has been used to foster social inclusion in the city. The case studies have been chosen to demonstrate the range of issues surrounding inclusion and public art. No claim is being made that the selection is either definitive or wholly representative. Due to the lack of evaluations into the success, or otherwise, of public art projects, it is difficult to compose a representative selection of good or bad practice and whereas iconic or controversial projects may receive critical and media attention, those at community level are often neglected in this respect. There are many examples of public art and it is difficult to choose examples without appearing anecdotal. Therefore, a deliberate attempt has been to consider a range of known works from Europe and North America many of which, in process as well as product, have become very influential contributing to subsequent debates surrounding notions of inclusion. That the works discussed are in a number of cases relatively well-known examples of public art should not be taken to imply that they necessarily reflect what might, in social inclusion terms, be considered as exemplars of good practice. Rather, their selection has been made to demonstrate the variety of public artworks and of the modes through which intervention can be sought. Further, the focus on process allows us to investigate how inclusive are practices, emphasising the key issue of ownership. Studies elsewhere on community participation have underlined the significance of ownership as shaping the value in which democratic participation is held (in the different case of housing, see Goodlad et al., 2001). Such a sense is apparent within the different 'stages' through which participation takes places from agenda setting to policy formulation to implementation, the critical factor being the extent to which, and how, citizens are included in the processes.

Public Art Production and Cultural (In)justice

1. Non-recognition: Reclaiming Place and Recognising Past

When working on participatory projects, artists are frequently dealing with communities who have been marginalised in mainstream urban histories. There is a general sense that they have been made invisible within the cityscape and therefore a key strategy in overcoming this sense of non-recognition is to render their history visible in some form. The very visibility of public art and its traditional monumentalism and aggrandising of civic 'heroes' mean that it is a prime vehicle through which minority groups can affirm their history and physically mark their place within the layered histories of the urban space—the past being a keystone upon which to build for the present and future. As Ron Griffiths has noted

> an important part of the experience of exclusion is a weakened or non-existent sense of identity and pride. A key step in integrating excluded populations into the social mainstream, therefore, is to assist them to find their voice, to validate their particular histories and traditions, to establish a collective identity, to give expression to their experiences and aspirations, to build self-confidence. (Griffiths, 1999, pp. 463–464).

There is a pervasive trend when working on such participatory projects to seek to, as Griffiths termed it, "validate their particular histories". Often the recognition of a particular community and their association with a specific place are integral to this process. Early and influential work in this area was undertaken by the non-profit arts organisation The Power of Place, which aims to create memorials or presences in the urban landscape of Los Angeles to those 'forgotten' by dominant histories—for example, the Black slave and midwife Biddy Mason (Hayden, 1995, pp. 169–187; Miles, 1997, pp. 177–178). Commissioned by the Community Redevelopment Agency and with funding from various organisations including the National Endowment for the Arts, The Power of Place liaised with the community and produced books, posters, a photomural and a 'Window of Memories' showing Betye Saar's *Nostalgia* collages. In addition, Sheila Levrant de Bretteville produced a permanent installation *Biddy Mason: Time and Place* on the site of Mason's homestead. Such modest interventions into public space restore and pay homage to the dignity of minority figures and yet, perhaps inevitably, are tainted with an air of nostalgia just as other monuments resound with patriarchal hegemony.

With some projects of this nature, there is an air of opposition, of reinstatement, of stressing an alternative canon when, perhaps, there is a need to yield more towards a 'differencing the canon', as the art historian Griselda Pollock has termed it (Pollock, 1999). As Dorothy Rowe explains

> Pollock implies that 'differencing the canon' is not about the replacement of one set of canonical works by another as devised by feminism, but that it is a rather more nuanced activity that continually questions the borders of knowledge, desires and power (Rowe, 2003, p. 28).

There is a danger amongst schemes that aim to resurrect tangibly histories that they will iconicise or nostalgically myth-make in a retelling of history. Against this, the nature of the contemporary community requires careful consideration, necessitating a questioning of its relationship with a particular place and its links, if any, with the past. As Lucy Lippard has stated

> Like the places they inhabit, communities are bumpily layered and mixed, exposing hybrid stories that cannot be seen in a linear fashion (Lippard, 1997, p. 24).

Replacing non-recognition with recognition seems somewhat simplistic and making the invisible visible, too literal. Although through its sheer visibility public art seems an ideal tool through which to restate a presence in the urban landscape and, by association, its history and evolution, it is here that the artist has the potential to intervene, to interact with the contemporary community, to research and reveal the past in a subtle and intuitive manner. In this, The Power of Place has sought to insert itself sensitively, creating varied forms of artwork and subtle inscriptions in the cityscape. To see its work as insular examples is to undermine the ethos of the project. Besides *Biddy Mason*, other projects have included the preservation of historic buildings in the 'Little Tokyo' district of the city (Hayden, 1995, pp. 210–225) and such places create a dialogue through which a sense of the urban experiences of minority communities can be felt.

Although a monument or memorial to a significant but neglected historical figure can have wide resonances for a minority community and its recognition at large, working with a collective history poses a different challenge. Recovering a neglected collective history that has little or no presence in hegemonic histories or traditional museum archives almost forces an artist to take an ingenious approach. The onus is specifically on the communal, the mutual endeavours and the shared struggles. To commemorate just one individual would undermine the *raison d'être* of the project. One innovative response to such a situation was demonstrated in Andrew Leicester's state-funded project *Prospect V-III* (1982) in Frostburg State College, Maryland. Again emphasising the importance of process, Leicester spent time visiting mining sites and interviewing miners and their families. The resultant work took the form of what could be termed an alternative kind of museum, built in the style of 19th-century mining architecture and housing artefacts donated by the miners' families. Furthering their involvement, members of the local community chose to provide guided tours, which suggested that they had taken 'ownership' of the work. As well as reinstating a presence in the landscape and recovering a lost history, Leicester made the project relevant for the contemporary community. Leicester's inclusive process and the resultant artwork have the potential to function as inspiration for those wishing to remember excluded communities in contemporary post-industrial cities.

In dealing with public art, it is tempting to focus on place-specific works as public art tends to be associated with particular places and situated in certain sites. However, just as notions of community can transcend specific geographical locales, so can artworks themselves. In 1991, the V&A initiated the *Shaimiana: The Mughal Tent* programme devised by Shireen Akbar and aimed at south Asian women and their children. Although encompassing broader aims than just recognition, Akbar's project saw inclusion as a challenge facing society at large and also, sensitively, as an internal cultural issue as Asian children were perceived to be losing a sense of their south Asian heritage as they became integrated into British culture. Community groups from across Britain were brought to the V&A to examine aspects of south Asian history using the museum's collection. As a result, they produced a textile panel using traditional methods but reflecting contemporary cultural concerns. The panel was then displayed as part of a tent, inherent in which were ideas of home, transience and travelling (Rowe, 2003). As well as opening the museum's collection to a wider community, the project brought together an immigrant ethnic group perceived as being isolated from mainstream British culture. Partly to help overcome this sense of isolation and to create dialogue between different female communities and cultures, those involved included non-Asian women and the project integrated a variety of religions. Therefore the project attempted to build connections across a marginalised group and cultivate relationships and awareness with other sections of society. The *Mughal Tent* became an international project and has been made accessible via the Internet, its impact rippling out to a much wider community altogether and its form challenging traditional conceptions of public art.

All of these artworks were inclusive in terms of their target audience and, crucially, their practice. They addressed communities that tended to be excluded from wider urban processes and, with them, created a tangible,

if, in the case of *The Mughal Tent* mobile, marker of their presence. However, this is not an overt claim for recognition through mere visibility. They overcome Fraser's notion of non-recognition in a tangible but subtle sense, trying to increase awareness of marginalisation and commemorate histories in a manner meaningful for the present. In this, a meaningful, democratic process has been key to the sustainability of initiatives and apparent in the outcomes.

2. Disrespect: Giving Voice, Countering the Stereotype and Rediscovering the Margins

If giving voice through the vehicle of public art can be the means of drawing the invisible into the urban narrative, it also has a role in drawing in those citizens and spaces whose marginalisation stems from other causes. That is, marginalisation in the city is not just a product of being invisible. The poor, those living in deprived neighbourhoods, are not so much invisible as inaudible. Stigmatisation, the stereotyping of particular groups and the urban spaces they occupy, is a commonplace source of marginalisation.

The idea of giving a community a voice and overcoming preconceptions was at the core of Iñigo Manglane Ovalle's work *Tele-Vecindario* (1992–93) for 'Culture in Action', an outreach project devised by the non-profit arts organisation Sculpture Chicago (since merged into Chicago Department of Cultural Affairs' public art section). Curated by the influential curator and writer, Mary Jane Jacobs, 'Culture in Action' was a deliberate attempt to engage minority communities unfamiliar with Sculpture Chicago and the artworld at large. Rather than artists with an international reputation, those chosen were known for participatory projects and, in this instance, went through a two-year period of collaboration with minority groups (Drake, 1994, p. 13). The resultant artworks took various forms from an altered paint chart to a multi-ethnic festival to sculpture and chocolate bars. Despite the length of collaboration, criticism has been made that artists tended to be 'shipped in', therefore having little knowledge of the communities with which they were working (Hixson, 1998), and the value of the projects has been questioned (Karasov, 1996, p. 25). Still, three of the eight artists involved were Chicago-based, including Ovalle whose project was arguably the most successful in the sense of engaging with the local community.

Ovalle lives and works in the lower-income, mainly Hispanic neighbourhood of West Town, an area riven by gang violence and the problems associated with social deprivation. Working with a community leader and a co-ordinator of a Schools Programme, Ovalle brought together a group of mostly Latino teenagers, some from rival gangs. It emerged that the teenagers felt misrepresented and stereotyped by the media and countering this became the impetus for their project. With the assistance of local video professionals, they formed Street-Level Video and produced a series of films of people in the community discussing a range of issues including gentrification, race and gangs. This not only formed links between various factions of the community but also bridged generations. The project culminated in a 'block party' where 71 monitors were placed outdoors and broadcast these dialogues to the public at large. Local residents voluntarily provided electricity from their homes and this wittily and symbolically furthered the notions of empowerment central to the project (Jacobs, 1995, p. 86). The party brought people from across Chicago into the neighbourhood, challenged preconceptions, and places of street violence became more neutral spaces for exchange and dialogue. Importantly, Ovalle's project was not simply a means to counter stereotypes and offer an alternative but homogenised narrative; instead, it involved a more perceptive communication of numerous identities within a community and encouraged dialogue both internally and with wider society.

A clear sign of the achievement of the project has been its sustainability; Street Level Youth Media remains active and the 'block party' is an annual feature. It is a moot point, however, whether this success is attributable to Ovalle's position as resident-artist. Ovalle inevitably benefited from pre-existing connections and knowledge and yet, as Jacobs herself has noted, it is perhaps oversimplistic to presume that "you're making a significant work for a place by solely selecting people who are full-time residents of that place" (Drake, 1994, p. 13). It is possible to argue that bringing an outsider's point of view, a fresh pair of eyes free from any preconceptions, would be equally beneficial. Therefore, emphasis must, again, fall on the implementation of a process, which, in this instance, was democratic giving room for multiple voices to be heard.

The articulation of numerous identities in an artwork, however, can be problematic. Although *Tele-Vecindario* recognised the various facets of one community, mainly Hispanic, process becomes more complex when seeking to implement a democratic practice and produce an inclusive artwork within an extremely mixed neighbourhood. In such circumstances, one solution has been to facilitate a space for cultural exchange rather than impose an artificial, fixed vision of a community through a singular representation. When there is a shared sense of history, religion, nationality or even loss contributing to a sense of community, then it is perhaps easier to find a common theme or direction for an artwork than when a community is multicultural and diverse, and whose links lie in their residency on the same site and their situation as a minority within the larger social fabric of the city. In such a social context, claims that public art can contribute to, if not create, community cohesion, seem somewhat 'easier said than done' and misguided. As Ash Amin has argued

The distinctive feature of mixed neighbourhoods is that they are communities without community, each marked by multiple and hybrid affiliations of varying social and geographical reach, and each intersecting momentarily (or not) with another one for common local resources and amenities. They are not homogeneous or primarily place-based communities ... They are simply mixtures of social groups with varying intensities of local affiliation, varying reasons for local attachment, and varying values and cultural practices ... Mixed neighbourhoods need to be accepted as the spatially open, culturally heterogeneous, and socially variegated spaces that they are, not imagined as future cohesive or integrated communities (Amin, 2002, p. 972).

Echoing the sentiments of Hall and Robertson mentioned above, Amin argues that difference must be an integral part of the process towards inclusiveness. Recognising this, the arts organisation nva worked with the most ethnically diverse community in Scotland in Glasgow's Pollokshields district to transform derelict wasteland into 'The Hidden Gardens'. The gardens, in planting, planning and a series of artworks and events, aimed to reflect the horticultural influences of the various faith communities in the local area. Crucially, it aimed to be inclusive throughout from initial conception to the final realisation and subsequently (in its on-going management). It was a delicate and difficult process that involved dialogue with a number of communities. Key here, particularly in the process of initiating dialogue between the multiple communities, was the appointment of a community facilitator sympathetic to the needs of local groups. After learning about the area and its constituent communities, drafts for the project were successively discussed between groups and the artists and horticulturalists hired by nva (a consultancy firm specialising in the field). The participatory process was critical in which local groups were encouraged to assume that participation would influence the design and use of the space. In other words, by emphasising that participation would help to define the nature of the proposal, its implementation and, over the longer term, its running, the dialogue sought to encourage a sense of ownership and empowerment (Paddison and Sharp, 2003, p. 11). Although the gardens have restricted opening hours and this may raise questions as to their accessibility as a public space, since their opening they have proved a popular space of retreat in an area underprovided with open spaces. They have also become a tourist landmark, a positive development in bringing visitors to a part of the city otherwise off the tourist map, and for encouraging awareness of the ethnic diversity of the city. Yet, by being drawn into the rhetoric of the city's marketing promotion—multiculturalism as evidence of the city's promotional adage as 'The Friendly City'—the development courts usurping the communities' sense of ownership. By this token, social inclusion is not just an aim of urban regeneration; it can become also a means for projecting the city's image.

The increasing importance given to community liaison in such permanent projects requires careful consideration: there can be a substantial difference between consultation and inclusionary practices fostering empowerment (Burns et al., 1994; Young, 2000). The difference between seeking an opinion, which might be little more than a public relations exercise, and involving communities more fundamentally in the deliberative process culminating in decision-making itself can have profound impacts on the creation and reception of a work and the community at large. One example is Portsmouth City Council's commissioning of Peter Dunn, most famous for his association with the Art of Change, to produce an art scheme for a sports and community facility in a deprived neighbourhood. Dunn collaborated with the community, other artists, architects and landscapers to transform the environment based on the principles of Agenda 21. The site was given unique identity through the design of earthworks, boundary walls, narrative pathways and wall hangings and its distinctiveness was underscored by the landmark sculptural work the *Wymering Tree*. Perhaps more significant than the development itself, however, was the community involvement. A community board took control of the management of the project so they were not only given a voice in the aesthetic nature of the site, but also played an integral and decisive role in the process and the implementation of strategies. The community involvement in both the artistic and policy processes of the project demonstrates the potential for the community to determine the nature of the artwork produced and be integral to the changes taking place around them. This works well at community level, but the implementation of such a process city-wide, where public art is increasingly being used as a promotional tool, is more problematic.

The incorporation of major public art projects into regeneration schemes has become a key factor in rebranding a city's image, especially in post-industrial towns—in the UK, Glasgow, Birmingham and Gateshead are principal examples where culture, including public art, has been vaunted as a force in changing each city's fortunes. The *Angel of*

the North by Anthony Gormley has become iconic not only for Gateshead but also indicative of the power that public art can yield as a tool in changing the perception of the 'post-industrial' to the 'cultural' city. It is the figure-head for a scheme throughout the city that has used public art for a variety of purposes (see Figures 1–3).

The Riverside Sculpture Park has reclaimed a derelict industrial site and used it as a means to bring art to the public. Similarly, *Windy Nook* by Richard Cole is a prominent but sympathetic land art feature that has transformed a colliery slagheap into an environmental art-site to be used by the community. The public also encounters art in a series of schemes at the Queen Elizabeth Hospital that explore the potential of the arts in

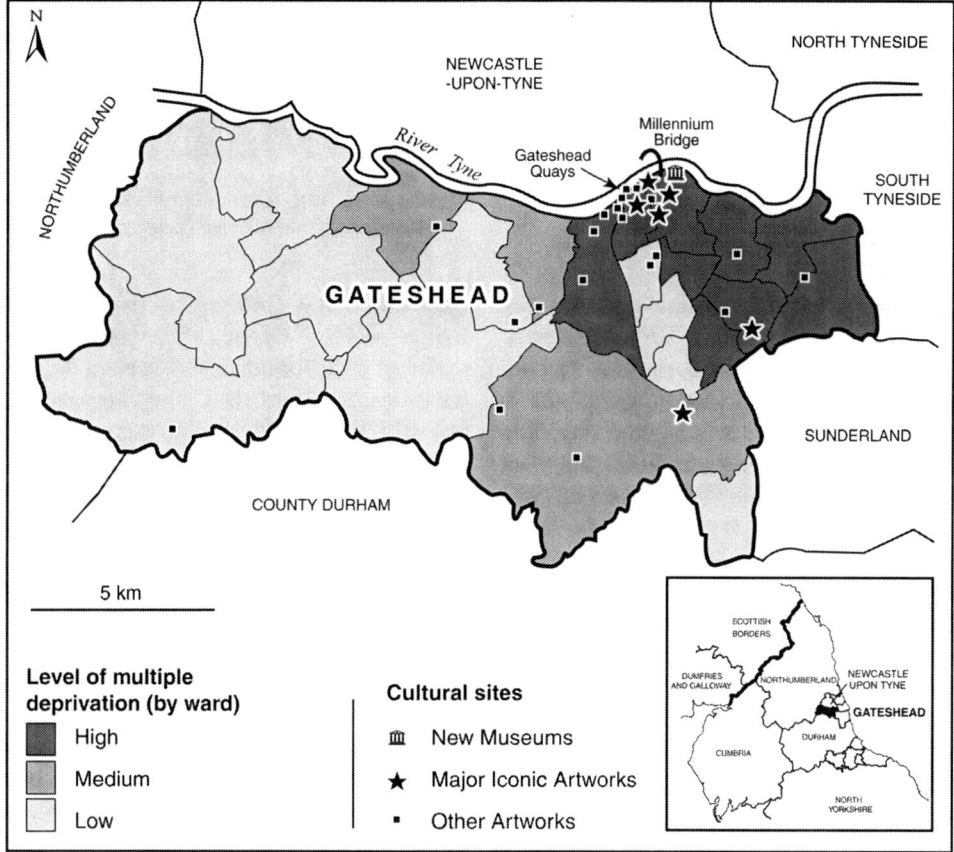

Figure 1. Public art and deprivation in Gateshead. Projects have focused on waterfront sites although these include several of the most deprived wards (defined on basis of socioeconomic and health indicators). The council has also sought to distribute public art works more widely in Gateshead.

Figure 2. Angel of the North (1998), by Antony Gormley. Photograph courtesy of Gateshead Council.

Figure 3. Rolling Moon (1998/90), by Colin Rose. Photograph courtesy of Gateshead Council.

healthcare. Alongside iconic pieces by renowned artists, including Richard Harris, Andy Goldsworthy, Colin Rose and Richard Deacon, the public is invited to participate in events such as the annual sculpture day. The iconic and community aspects are not wholly mutually exclusive, however, and over 1400 children at 30 schools were involved in work connected with Gateshead's *Angel*. Many of the artworks are concentrated in areas of social deprivation and this highlights how public art has been used as a tool to reaestheticise areas within a city as well as the city at large. Implicit in this is the notion that public art can bring economic and social benefits alongside the aesthetic. Their award-winning public art programme has evolved alongside several more iconic projects within the city centre including the conversion of old flourmills into the Baltic Arts Centre, the development of Gateshead

Quays and the building of the Millennium Bridge linking Gateshead's Quays to those recently refurbished in Newcastle. Together, the cities, which made a joint (but unsuccessful) bid for the 2008 European Capital of Culture nomination, aim to create a centre for cultural excellence with prestige residential and leisure developments. Here, the combination of iconic visual emblems, art and architecture, have gone some way to crafting the city as a cultural landscape.

It must be recognised that, although many policy documents and participatory projects appeal to overarching terms such as 'community', 'identity' and 'place', the general conformity of sentiment belies the complex situations facing artists, cities and their multifarious communities. The general lack of evaluative measures in community programmes means that it is difficult to outline measures of 'good practice', make affirmations of what constitutes a 'successful' intervention or add credence to the claims made about public art's social impact. As urban regeneration initiatives attempt to transform cities, public art itself seems to be undergoing a transformation, moving from traditional civic monumentalism towards seeking a more socially inclusive and aesthetically diverse practice. Rather than impose or enforce inclusion, it has to be intuitively and sensitively sought, recognising the importance of difference and the vitality in diversity. When working with inclusive projects, the emphasis seems to lie with the process rather than the product. This may raise questions about the artist deferring to the community, artistic integrity and aesthetic quality. It challenges institutions and funding bodies to consider the worth of intangible as well as tangible outcomes, temporary as well as permanent products. Traditional notions of the artist as creative genius have to be reassessed and the artist has to find a position between ingenious creator and creative facilitator. The use of culture in the reastheticisation of the urban environment also brings the danger that areas, as they become more attractive places in which to live and work, experience gentrification. Communities, places and process are integral but intricate components of the artwork. The extent to which each is considered and the manner in which their inclusion is sought and managed has profound implications for the aesthetic and social outcomes of the completed project.

3. Cultural Domination and the Arts of Resistance

Works such as Gormley's *Angel of the North* or monumental architecture that signal a city's distinctiveness (for example, London's 'gherkin' building, Glasgow's Scottish Exhibition and Conference Centre, Birmingham's mirrored Selfridges store or the competition between global cities to own the world's tallest building), demonstrate the importance of reworking the skyline as an attempt to refashion the image of the city as a whole. Such developments are clearly intended to enhance the image of the city, repackaging it as a commodity for consumption in the post-industrial age (see Urry, 2001). Public participation is generally not high on the agenda in this form of 'authoritarian populism', a Victorian image of city leaders knowing what's best for the city and for its citizens. In claiming to be a signifier for the city as a whole, of course, it hides the inclusions and exclusions inherent in *any* singular vision for a community.

Furthermore, the ever-increasing use of notions of inclusion through public art works in urban regeneration efforts makes one important assumption: that these processes are compatible. However, as Malcolm Miles has argued, due to the very nature of capitalist development, it is not always possible to draw public art into development in a way that is equally beneficial to all parties

> Developers do not develop in order to construct the 'city beautiful'. They construct the city beautiful in order to conceal the incompatibility of their development with a free society (Miles, 1997, p. 130).

From such a perspective, any sense of involvement with the process then is inherently linked to collusion with forces that are fundamentally more interested in capital investment or maintaining social order than with improving the lives of residents of a city. Certainly, critics of Glasgow 1990 City of Culture were quick to point out the benefits to industry and investment in the city but that, for the majority of Glaswegians, nothing had changed (Boyle and Hughes, 1991). Not dissimilar criticisms have arisen in other cities as Broudehoux (2004) has demonstrated in *The Making and Selling of Post-Mao Beijing*. For others, there is a problematic of where public space is—in contemporary cities, the spaces where people meet are increasingly being commodified so making truly grassroots expressions in public space more difficult (Mitchell, 2003; Phillips, 1988). Therefore, it is debatable whether public art can ever be wholly inclusive, especially within urban regeneration where complex factors of public space, commercialisation and commodification, and cultivating an iconic cultural cityscape are intimately entwined. In this environment, public art, rather than participating in an inclusive agenda, can function as an oppositional or resistant force, highlighting excluded groups and visualising protest to dominant regeneration schemes.

4. Resistance and Regeneration

As opposed to the usual iconic permanence of art prominent in regeneration schemes, these interventions tend to be temporary and take a variety of forms.

For instance, the redevelopment of the London Docklands in the 1980s stimulated a battle for land and for visibility that was articulated through the nature of the visual landscape in art. The reimaging of the landscape, however, neither accepted nor accommodated all: local people felt that not only had they been dispossessed by the new developments but that they had been written out of this new landscape. Developers talked in terms of a 'virgin site' for development, erasing the resident population and those who had lived there in the past. The gentrified landscape romanticised a particular part of the area's history, focusing on middle-class cultures of consumption rather than working-class cultures of production from the area (just as was later to occur in Glasgow where the rewriting of the city's landscape in terms of Charles Rennie Mackintosh and the Glasgow School of Art were seen to be at the expense of working-class cultures of 'Red Clydeside' and social struggle (Boyle and Hughes, 1991)). In addition, the new developments privatised space: waterfront walks became private property, high walls were erected around new development and the Docklands Highway ran through the Trade Unions building and local housing. One of the attempts to rewrite silenced community back into landscape was through the Docklands Community Poster Project which visualised narratives of community presence back into the redeveloped landscape through the use of billboards showing aspects of local history and identity, and expressing people's opinions of the development (Dunn and Leeson, 1993; Bird, 1993). These large posters displayed local versions of place in the landscape of regeneration, refusing to be hidden behind the reworked image. They aimed to write people and their history into the landscape rather than an aesthetic focused around property and heritage.

Other expressions of resistance to this redevelopment took the form of vandalism and graffiti. To some, as a form of cultural expression, graffiti are giving a voice to those who do not own the capital and buildings upon which billboards are mounted, and so cannot legitimately write up their messages onto the urban landscape. It is a way in which those who have been passed over by regeneration can write themselves back into the landscape, refusing to conform to the new urban order. Some feel that, although we are "as much creatures of the public realm as the private realm, we find ourselves silenced whenever and wherever we might create meaning to share with others" as public space is tightly controlled by capitalism and the capitalist state (Luna, 1995, np). *Graffiti* and culture jamming (the addition of slogans to billboards and advertising to subvert the intended message) draw attention to the power and meanings inscribed into the urban environment. Its artists are attempting to denaturalise the taken-for-granted landscapes that we each use on a daily basis, asking us to be aware of the power relations that work through this mundane space (Cresswell, 1998; Deutsche, 1996).

Some have incorporated this politics of opposition into their approach to creating art in the city—for instance, Krystof Wodiczko, particularly in his series of *Projections*. These nighttime projections of images onto prominent buildings and statues were used to challenge the meanings of the landscape elements that dominate contemporary cities. Thus, in projecting a swastika onto the South African embassy in London 1985, missiles onto war memorial columns and images of disability onto heroic statues, Wodiczko was asking the viewer to think about what it is we choose to memorialise in our landscapes, and which experiences, events and narrations of history are silenced (Wodiczko, 1999).

Artists such as Wodiczko also seek to represent those in the urban landscape who cannot represent themselves. As the Docklands example showed, there are many who are marginalised by processes of redevelopment, or who are displaced by the processes of regeneration (Smith, 1993). Wodiczko developed the *Homeless Vehicle Project*, a mobile vehicle for homeless people to use to sleep, wash and keep their belongings in. It made homeless people visible, drawing them out from their naturalisation as an accepted (if unfortunate) part of the modern urban landscape (Smith, 1993). The vehicle became a talking-point between pedestrians and the homeless, and its consciously missile-like design made comparisons between the US government's spending on social welfare and defence unavoidable. Michael Rakowitz similarly exposed the relationship of homelessness to urban development in his *ParaSITE* series which placed inflatable constructions over heat vents of buildings to provide homeless people insulated and private spaces where they could exist—parasitically—alongside the modern buildings. As he put it

> While these shelters were being used, they functioned not only as a temporary place of retreat, but also as a station of dissent and empowerment; many of the homeless users regarded their shelters as a protest device ... The shelters communicated a refusal to surrender, and made more visible the unacceptable circumstances of homeless life within the city (Rakowitz, 2000, pp. 234–235).

Clearly, the homeless vehicle project and *ParaSITE* are not solutions to homelessness and the other negative social products of urban redevelopment. Their role as a cultural product of the city is to make visible the naturalised relationships that are established through the built form, to amplify "the problematic relationship between those who have homes and those who do not have homes" (Rakowitz, 2000, p. 235). As Fraser argued,

amongst the debates surrounding the use of culture in urban regeneration attention needs to be paid to economic redistribution. Works like those of Wodiczko and Rakowitz highlight the problematic notions of inclusion in schemes of urban redevelopment.

5. Problems of Process

In practice, the division between public art initiated by authorities and grassroots approaches is much messier than it may seem. State-sponsored projects now almost always include a community element, while some artists who have previously worked with grassroots projects now work for both 'top–down' and community-led projects (for instance, the *Art of Change*). Critical artists claim that their work establishes a conversation between the spaces and the people who inhabit them. This is perhaps questionable: who really has a conversation? To what extent does this rely upon an élitist language of art and politics? On the other hand, art that is developed through the effort of local governments and other local development agencies does not necessarily turn out the way that was intended, alternative meanings and practices might emerge, people can reinscribe images with personal and local meaning. As mentioned earlier, this makes clear the importance of understanding the *processes* through which public art is made and placed within different parts of the city. When this fails, it can promote interesting debates over issues, but can also be intensely demoralising for the communities involved.

The complexity of the processes through which public art is made meaningful to different communities is perhaps most (in)famously seen in debates over Richard Serra's *Tilted Arc* (see Senie, 2002). The work was sited in Federal Plaza, Manhattan in 1981 as one of the last in the Kennedy-inspired Art-in-Architecture programme to bring art into public spaces. Serra, an artist known for his 'anti-environmental' works, saw his sculpture as challenging the bourgeois bureaucratic spaces that usually contextualise the display of artwork—in this case, the sanitised, alienating square created by the meeting of two blocks of the Federal Building. The sculpture was constructed from Corten Steel, 120 feet long, 12 feet high and 2.5 inches in width. Covered with a surface of brown rust, *Tilted Arc* bisected the square, tilting off both its horizontal and vertical axes (Blake, 1993, p. 261). Serra challenged this order through a sculptural form that refused to offer a reconciliation of architecture and sculpture but instead revealed "a conflicted space that lays bare its internal divisions to its inhabitants" (Blake, 1993, p. 254).

There was a great deal of opposition to the sculpture. Some saw it as too oppressive, too big, too dominant or too rusty. New Reaganite federal leaders used this popular opposition to push for the removal of the artwork as part of an attack on the National Endowment for the Arts and radical art more generally. These right-wing opponents assumed that public opposition meant that, like them, the public rejected *Tilted Arc* in favour of the previous environment of the square. This was not quite the case. On the whole, it appeared that public opinion was in agreement with Serra's critique of the alienating square but not the aesthetic form he had adopted. They found the sculpture as sterile as the space it sought to subvert. One worker said "I do not care to be challenged on a daily basis by something designed to be hostile" and another concluded that "What we need . . . is something to enliven our lives, not something which reinforces the negativity of our work lives" (Blake, 1993, p. 284).

This raises important questions about artists' responsibility to community. Some of the most artistically successful and challenging work may not be easy to live with. While it is possible to walk away from a work in a gallery, once works are incorporated into lived spaces they cannot always be avoided.

What the *Tilted Arc* controversy forced us to consider is whether art that is centered on notions of pure freedom and radical autonomy and subsequently inserted into the public sphere without any regard for the relationship it has to other people, to the community, or *any* consideration except the pursuit of art, can contribute to the common good (Gablik, 1995; quoted in Miles, 1997, p. 90).

For post-*Tilted Arc* work, there has been less in the way of 'parachuted in' artists (but see Public Art Review Special (July/August 1998) *Public Art: Fail*) and, instead, context and community involvement are increasingly important. However, this is not to say that these tensions have disappeared (regardless of how inclusive the intentions are).

Furthermore, while properly managed processes can help to maximise a sense of ownership and even empowerment, if these processes are interrupted for whatever reason this can have negative consequences for the communities involved and for future attempts at community participation. An example of this was the *Five Spaces*, public spaces developed by artists and architects with communities around Glasgow as part of *Glasgow 1999 (UK City of Architecture and Design)*. Although the *Five Spaces* were to be one of the 'flagship' events, when 1999 arrived they enjoyed a much less significant public profile than the other spectacular (and centralised) events, Homes for the Future and the design centre the Lighthouse. While the Lighthouse took nearly half of *Glasgow 1999*'s budget of around £27.5 million, the *Five Spaces* was allocated less than one-tenth of it. Media coverage of the *Spaces* was similarly less prominent than for the Lighthouse and Homes for the Future, so that, when asked at the end of the year, very few visitors, Glaswegians or even design professionals could name the *Five Spaces* as a prominent feature of the programme (DTZ Pieda, 2000).

A 'trial run' of spaces was implemented in 1997, when the artists took up their residencies and developed plans for the spaces. This aimed to develop a process through which communities could become involved in the selection of the spaces and the kind of work to be included. Each of the designated communities was located around the city, often in challenging environments. For some, the excitement of the *Five Spaces* was that, for the first time, the communities themselves were being asked for what they wanted rather than 'experts' telling them what they needed. Process was central to this project to ensure that there was community involvement—and therefore hopefully a sense of ownership—in the resulting spaces. Initially, there were to be over 20 of these spaces chosen by Housing Associations around the city. The initial cut took the number to 15, then to 11 and then, well into 1998, the number was reduced to 5. For those who had been expecting their space to be developed, this was a major blow and perhaps reinforcement of the sense that their community was marginal to the city. Even for those communities which did have their spaces developed, the culture of uncertainty and experiences of being let down before meant that community leaders were unwilling to involve the community until the funding was absolutely assured by which point there was not sufficient time for proper participation.

Glasgow 1999 were determined that all the spaces would be delivered in 1999 (all but one were) and so they put in place a property management firm to deal with the arrangements of making the space. Here, institutional power heavily influenced the process–governance framework. The day-to-day ownership of the project was taken away from the Housing Associations for the sake of efficiency. Many Housing Association members felt that this pushed them out of the decision-making process and there has consequently been a loss of ownership, which has impacted the maintenance and management of the *Spaces*. Long-term problems are emerging because of this interruption of the process and lack of consideration to sustaining the *Spaces*— *Glasgow 1999* did its job and disbanded whereas the *Spaces* remain. The Housing Associations were not consulted about small, everyday issues and consequently, in one instance, the landscaping in one of the *Spaces* has been easily vandalised. Two years' labour was arranged for upkeep, but now it is up to the Associations to make arrangements, which it appears some cannot

manage, or, as a result of their exclusion from the decision-making process, choose not to.

A great deal of the good that was done through these projects—of bringing people in to feel a sense of communal ownership, of making networks and so on—has been undone. Harding explains the problems that emerge when acts of vandalism are not immediately righted—and here we could also add other forms of decay such as flooding, breaking of light bulbs and problems with water features, all issues plaguing the *Five Spaces*.

> When this happens, what was initially a focus of local pride quickly degenerates to the point where people become even more disheartened than they were before. Rectifying the damage done by vandals immediately sends out a clear message to people in deprived areas that their welfare is just as important to the authorities as the well-being of people living in affluent circumstances (Harding; quoted in Gordon, 2002, np).

However, leaving things in decline reinforces the image of a community in similar trouble. The story of the *Five Spaces* emphasises the importance of good process for the success of public art, but also makes clear just how fragile process can be, that it is "seemingly capable of derailment at any juncture for a variety of reasons" (Nikitin, 2000, np).

The infamy of the *Tilted Arc* controversy, and others like it, has meant that community involvement and consultation are now central to the process of siting and producing works. However, this does not mean that inclusion is necessarily achieved. Quite how it is that 'the community' should be involved can, in itself, become an exclusionary practice. Often, the same members of the community become involved and consequently others may feel further marginalised in light of these activists' participation. There is also often a spatial and temporal essentialism in defining community. While city-centre art works are seen as serving the whole city, when attention turns to public art in more marginal areas, there is a sense of a number of unchanging and spatially discrete communities existing in a neat patchwork across the city. Inclusive approaches to designing public art—workshops, meetings and so on—can ensure ownership by those in that community and hence encourage care and lessen the likelihood of vandalism. However, as communities within cities are not water-tight spaces, this will not prohibit members of nearby neighbourhoods from (mis)using the public art too. Similarly, approaches that draw in community members at one time cannot ensure that future community members will still feel a sense of ownership over the product. As Senie (2003) suggests, this requires a critical rethinking of notions of site-specificity. She argues that as "a public site invariably undergoes seasonal and/or developmental changes, any work would logically have to be frequently or periodically redesigned to remain specific" (Senie, 2003, np). Once again, process, here defined in the long term, is central to success.

Perhaps too much is expected of public art. Too quickly, a number of critics have blamed these projects for not making enough of a difference. John Calcutt ridicules such expectations

> Expecting public art to solve social problems is either naïve or cynical. In attempting to critically evaluate public art projects such as *Five Spaces* we should bear in mind that fact that the production of art arises within and is subject to many of the same social, political and economic pressures that affect its reception (the increasing privatisation and commercialisation of the public sphere, the fragmentation of unified social and political agendas into the specialised concerns of competing interest groups—each with their own social and cultural priorities, and so on) (Calcutt, 2002, p. 11).

Calcutt is not suggesting that public art such as the *Five Spaces* is somehow put beyond criticism, but that it is impossible for such works to transcend their social, political and, perhaps most importantly in the case of urban regeneration, economic context. This again points to a

need for developing appropriate forms of evaluation. Rather than an evaluator's post-production critique, however, there is a clear need for evaluation to take place throughout the process—from the inception, through the process to the final work. Assessment needs to be made of the process and its success in being 'inclusive' as well as the governance structure through which it is implemented. Such evaluation has the potential to ensure that a meaningful process would yield a meaningful outcome and that problems of process are overcome.

Conclusions

In a critique of the uses of culture in the reinvention of Barcelona that is all the more refreshing precisely because of the frequency with which the city is cited as a role model, Balibrea (2001) has argued that the consensus over the city's development needs to be challenged. As significant as have been the achievements of the city, and particularly of the municipal government, in physically transforming the metropolitan area including run-down inner-city areas as well as the waterfront and harbour, and in bolstering its economic competitiveness, particularly as a tourist and convention centre, such achievements have been accompanied by increasing social polarisation and the development of peripheral estates whose residents have endured a worsening quality of life in both the 1980s and 1990s. The absence of any significant dissent ('culture wars') particularly over flagship projects may be read as support for change. But it may also be read as false consciousness in which "the production of consensus [as] the principal means of legitimising domination and of co-opting potentially critical citizens" has been able to convince the citizens that their interests are equivalent to those of dominant economic classes (Ripalda, 1999; Esquirol, 1998; quoted in Balibrea, 2001).

While such alternative interpretations can themselves be challenged, their value lies in unmasking the rhetoric that surrounds the use of culture—including the benefits claimed for public art—in urban regeneration. Here, the 'Barcelona model' represents something of an extreme precisely because of the frequency with which it is cited, although exemplars of 'good practice' elsewhere are routinely identified as 'success stories'. For urban regeneration agencies, the search to repeat the 'Guggenheim effect' has become a mantra through which the reinvention of the city is to be realised within which public art, and particularly iconic design, occupies a critical position.

As the literature attests, it is too easy for both policy-makers *and* academics to focus disproportionately on the more spectacular, particularly the iconic, in its ability to reinscribe place. A blinkered gaze risks the failure to identify the different scales at which public art has come into play just as it tends to give emphasis to particular representations of it. As various examples have demonstrated, the use of public art is no more confined to the major cities as it is only to those spaces in them whose (re)valorisation has become part of the hegemonic project of fashioning the competitive city. In the interstices—in those places and spaces which are 'outside' the dominant discourse of *international* competitiveness that characterises the big city—the recognition of the contribution of public art to the reinscription of *local* place has become commonplace through the work of artists and community groups, as well as by the state acting through local agencies mindful of the agenda of inclusion.

It is important to remember here that, regardless of the scale and type of intervention, the installation of public art within the urban fabric is inevitably a political exercise. Thus, as Jameson has argued, buildings

> interpellate me—[they] propose an identity for me, an identity that can make me uncomfortable or on the contrary obscenely complacent (Jameson, 1997, p. 129).

Much the same could be said for public art. The roots of this effect lie in its visibility, which in turn influences how we perceive the urban environment. Admittedly the influence of agency at this juncture means that how we perceive and interpret the interposition of public art varies as Jameson recognises and as was apparent through the Donald Dewar statue. Inevitably reactions to public art will vary. But even amongst those whose reaction to public art is more passive or 'complacent', its effects on the definition of the self and the self as citizen are real, if unarticulated.

The power of interpellation of public art is both a source for consensus and conflict within the reinscription of place. Within official discourse, the benefits of public art are expressed in its ability to instil civic pride and to contribute to local distinctiveness, yet the ability of public art to be seen as at odds with its intended symbolism emphasises its contentious nature. The play of inclusion in public art operates at two interconnected levels in the ways in which it is read as part of city space and the processes through which it is implemented. Sufficient experience exists to demonstrate that the two are connected, suggesting that a sense of ownership is a key component of inclusion. Yet neither is fixed precisely because of the multiplicity of ways in which public art is read and the fluidity of urban societies that defy the unity of community.

As much as this is suggestive of the importance of participation within the process of the production of public art, its advocacy will need to take account of the problems typically encountered in its practice. How local participation is structured to give adequate recognition to different local groups and how deliberation is conducted to ensure that these interests are able to have their voice heard and listened to are both fundamental to the practice of inclusive democratic processes. In both, experience of local participation highlights the problems likely to arise: the extent to which it can be dominated by a relatively small number of local activists, the reluctance to become involved and the problems in ensuring that meetings are conducted on the basis of equality and mutual respect and recognition.

Even, then, in the interstitial banal spaces in which everyday life is locally lived within cities, the installation of public art needs to be sensitive to local diversity. Its use needs to be aware that inclusive democratic practices, far from producing consensus—through which some common sense of local pride can be produced amongst diverse groups—may become an agonistic process. The exercise of participatory democracy through recognition and respect and the avoidance of domination opens up the space for conflict reflecting the diversity of local voice. The reinsertion of public art in the city reveals how its use, and the language through which it is advocated, can be appropriated, no more so than in those revalorised spaces that are identified as key to the (re)definition of the competitive city. It is in the banal urban spaces in which everyday life is constructed and experienced in particular that the advocates of public art have been able to argue (and demonstrate) how the insertion of public art can aspire to be inclusive as process if not necessarily as outcome.

Yet the capacity of public art to foster inclusion is at best partial, able to address symbolic more than it is material needs. Whether this means that public art has become an unwitting agent in the overprivileging of cultural justice at the expense of socioeconomic redistribution is a moot point. However, this argument not only exaggerates the influence of public art on economic regeneration, but is itself an overeconomistic interpretation of the meaning of urban citizenship.

References

AMERICANS FOR THE ARTS (ND) Animating Democracy Initiative: Profiles and Cast Studies—Tele-Vecindario (http://www.americansforthearts.org/animatingdemocracy/resource_center/profiles_content.asp?id = 185; accessed 28 June 2004).

AMIN, A. (2002) Ethnicity and the multicultural city: living with diversity, *Environment and Planning A*, 34, pp. 959–980.

ANDERSEN, J. and SIIM, B. (Eds) (2004) *The Politics of Inclusion and Empowerment: Gender, Class and Citizenship*. Basingstoke: Palgrave Macmillan.

BALIBREA, M. P. (2001) Urbanism, culture and the post-industrial city: challenging the 'Barcelona Model', *Journal of Spanish Cultural Studies*, 2(2), pp. 187–210.

BIANCHINI, F. (1999) Cultural planning for urban sustainability, in: L. NYSTROM (Ed.) *City and Culture: Cultural Processes and Urban Sustainability*, pp. 34–51. Stockholm: The Swedish Urban Environment Council.

BIRD, J. (1993) Dystopia on the Thames, in: J. BIRD, B. CURTIS, T. PUTNAM *ET AL.* (Eds) *Mapping the Futures*, pp. 120–135. London: Routledge.

BLAKE, C. (1993) An atmosphere of effrontery: Richard Serra, *Tiled Arc*, and the crisis of public art, in: R. FOX and T. J. LEARS (Eds) *The Power of Culture*, pp. 247–289. Chicago, IL: University of Chicago Press.

BODDY, M. and PARKINSON, M. (2004) Competitiveness, cohesion and urban governance, in: M. BODDY and M. PARKINSON (Eds) *City Matters: Competition, Cohesion and Urban Governance*, pp. 407–432. Bristol: Policy Press.

BOYLE, M. and HUGHES, G. (1991) The politics of the representation of 'the real': discourses from the Left on Glasgow's role as European City of Culture 1990, *Area*, 23(3), pp. 217–228.

BROUDEHOUX, A. (2004) *The Making and Selling of Post-Mao Bejing*. London: Routledge.

BURNS, D., HAMBLETON, R. and HOGGETT, P. (1994) *The Politics of Decentralisation: Revitalising Democracy*. London: Macmillan.

CALCUTT, J. (2002) Rack and ruin: the misplaced aims of public art, *Matters*, 15, p. 11.

CRESSWELL, T. (1998) Night discourse: producing/consuming meaning on the street, in: N. FYFE (Ed.) *Images of the City: Identity and Control in Public Space*, pp. 268–279. London: Routledge.

CZAPLICKA, J. J. and RUBLE, B. A. (Eds) (2003) *Composing Urban History and the Constitution of Civic Identities*. Washington, DC: Woodrow Wilson Center.

DEUTSCHE, R. (1996) *Evictions: Art and Spatial Politics*. Cambridge, MA: MIT Press.

DRAKE, N. (1994) Making it happen: a dialog with Mary Jane Jacob, *Public Art Review*, 10, 5(2), pp. 13–15.

DTZ PIEDA CONSULTING (2000) *Evaluation of Glasgow 1999: UK City of Architecture and Design*. Glasgow.

DUNN, P. and LEESON, L. (1993) The art of change in the Docklands, in: J. BIRD, B. CURTIS, T. PUTNAM *ET AL.* (Eds) *Mapping the Futures*, pp. 136–149. London: Routledge.

ESQUIROL, J. M. (1998) *La frivoldad politica del final de la historia*. Madrid: Caparros Editores.

FRASER, N. (1995) From redistribution to recognition? Dilemmas of justice in a post-socialist age, *New Left Review*, 212, pp. 69–93.

FRASER, N. (1997) *Justice Interruptus: Critical Reflections on the Post-socialist Condition.* London: Polity Press.

GABLIK, S. (1995) Connective aesthetics: art after individualism, in: S. LACY (Ed.) *Mapping the Terrain: New Genre Public Art.* Seattle, WA: Bay Press.

GIDDENS, A. (1998) *The Third Way.* Cambridge: Polity Press.

GOODLAD, R., DOCHERTY, I. and PADDISON, R. (2001) *Citizen participation in urban governance.* Policy Paper 4, Scottish Executive. Department of Urban Studies, University of Glasgow.

GORDON, G. (2002) When art goes public, *Scotland on Sunday* 16 June (http://news.scotsman.com/archive.cfm?id = 653692002; accessed 22 October 2002).

GRIFFITHS, R. (1999) Artists organisations and the recycling of urban space, in: L. NYSTRÖM (Ed.) *City and Culture: Cultural Processes and Urban Sustainability*, pp. 460–475. Karlskrona: Swedish Urban Environment Council.

GUNN, S. (2000) *The Public Culture of the Victorian Middle Class: Ritual and Authority and the English City.* Manchester: Manchester University Press.

HALL, T. and ROBERTSON, I. (2001) Public art and urban regeneration: advocacy, claims and critical debates, *Landscape Research*, 26(1), pp. 5–26.

HARVEY, D. (2000) *Spaces of Hope.* Edinburgh: Edinburgh University Press.

HAYDEN, D. (1995) *The Power of Place: Urban Landscapes as Public History.* Cambridge, MA: MIT Press.

HIXSON, K. (1998) Icons and interventions in Chicago and the potential of public art, *Sculpture Magazine*, 17(5) (www.sculpture.org/documents/scmag98/chicgo/sm-chcgo.htm; accessed 28 June 2004).

IMRIE, R. and RACO, M. (2003) *Urban Renaissance? New Labour, Community and Urban Policy.* Bristol: Policy Press.

JACOBS, M. J. (Ed.) (1995) *Sculpture Chicago: Culture in Action.* Seattle WA: Bay Press.

JAMESON, F. (1997) Absent totality, in: C. D. DAVIDSON (Ed.) *Anyone*, pp. 124–131. Cambridge, MA: MIT/Anyone Corporation.

KARASOV, D. (1996) Is placemaking art?, *Public Art Review*, 15, 8(1), pp. 24–25.

LACY, S. (Ed.) (1995), *Mapping the Terrain: New Genre Public Art.* Seattle, WA: Bay Press.

LEFEBVRE, H. (1991) *The Production of Space.* Oxford: Basil Blackwell.

LEVINSON, S. (1998) *Written in Stone: Public Monuments in Changing Societies.* Durham, NC: Duke University Press.

LEVITAS, R. (1998) *The Inclusive Society? Social Exclusion and New Labour.* Basingstoke: Palgrave Macmillan.

LIPPARD, L. (1997) *The Lure of the Local: Senses of Place in a Multicentred Society.* New York: New Press.

LUNA, J. (1995) Eradicating the stain: graffiti and our public spaces, *Bad Subjects*, 20 (April) (http://www.hiphop-network.com/articles/graffitiarticles/graffitiand publicspace.asp; accessed 28 June 2004).

MADANIPOUR, A., CARS, G. and ALLEN, J. (Eds) (1998) *Social Exclusion in European Cities: Processes, Experiences and Responses.* London: Jessica Kingsley.

MARCUSE, P. and KEMPEN, R. VAN (Eds) (2002) *Of States and Cities: The Partitioning of Urban Space.* Oxford: Oxford University Press.

MILES, M. (1997) *Art, Space and the City.* London: Routledge.

MITCHELL, D. (2000) *Cultural Geography: A Critical Introduction.* Oxford: Basil Blackwell.

MITCHELL, D. (2003) *The Right to the City.* New York: The Guilford Press.

MOULAERT, F. Swyngedouw, E and RODRIGUEZ, A. (2003) *The Globalized City.* Oxford: Oxford University Press.

NIKITIN, C. (2000) Making public art work, *Sculpture magazine online* (www.sculpture.org/documents.scmag00/april00/pub/pub.htm; accessed 27 July 2004).

PADDISON, R. and SHARP, J. (2003) *Towards democratic public spaces.* On-line papers, Department of Geography and Geomatics, University of Glasgow (http://web.geog.gla.ac.uk/online_papers/rpaddison001.pdf; accessed 27 July 04).

PHILLIPS, A. (2004) Identity politics: have we now had enough, in: J. ANDERSEN, and B. SLIM (Eds) *The Politics of Inclusion and Empowerment: Gender, Class and Citizenship*, pp. 36–48. Basingstoke: Palgrave Macmillan.

PHILLIPS, P. (1988) Out of order: the public art machine, *Artforum*, December, pp. 92–96.

PHILO, C. and KEARNS, G. (1993) *Selling Places.* Oxford: Pergamon Press.

PLAZA, B. (2000) Evaluating the influence of a large cultural artifact in the attraction of tourism: the Guggenheim Museum Bilbao case, *Urban Affairs Review*, 36(2), pp. 264–274.

POLLOCK, G. (1999) *Differencing the Canon.* London: Routledge.

RAKOWITZ, M. (2000) ParaSITE, in: J. HUGHES and S. SADLER (Eds) *Non-plan: Essays on Freedom Participation and Change in Modern Architecture and Urbanism*, pp. 232–235. Oxford: The Architectural Press.

RIPALDA, J. M. (1999) *Politicas Postmodernas: Cronicas desde la Zona Oscura.* Madrid: Libros de la Catarata.

ROWE, D. (2003) Differencing the city, in: M. MILES and T. HALL (Eds) *Urban Futures*, pp. 27–34. London: Routledge.

SASSEN, S. (2001) Cities in the global economy, in: R. PADDISON (Ed.) *Handbook of Urban Studies*, pp. 256–272. London: Sage.

SELWOOD, S. (1995) *The Benefits of Public Art*. London: Policy Studies Institute.

SENIE, H. (2002) *The* Tilted Arc *Controversy: Dangerous Precedent?* Minneapolis, MN: University of Minnesota Press.

SENIE, H. (2003) Responsible criticism: evaluating public art, *Sculpture*, 22(10), no page numbers.

SENIE, H. (2004) Absence in presence: the 9/11 memorial design, *Sculpture*, 23(4), no page numbers.

SMITH, N. (1993) Homeless/global: scaling places, in: J. BIRD, B. CURTIS, T. PUTNAM *ET AL.* (Eds) *Mapping the Futures*, pp. 87–119. London: Routledge.

SMITH, N. (1996) *The New Urban Frontier*. London: Routledge.

URRY, J. (2001) *The Tourist Gaze*. London: Sage.

VICTORIA AND ALBERT MUSEUM (ND) Shamania Microsite (http://www.vam.ac.uk/vastatic/ microsites/shamiana/voice.htm; accessed 13 July 2004).

WHELAN, Y. (2003) *Reinventing Modern Dublin: Streetscape, Iconography and the Politics of Identity*. Dublin: University College Dublin Press.

WODICZKO, K. (1999) *Critical Vehicles: Writings, Projects, Interviews*. Cambridge, MA: MIT Press.

YOUNG, I. M. (1990) *Justice and the Politics of Difference*. Princeton, NJ: Princeton University Press.

YOUNG, I. M. (2000) *Inclusion and Democracy*. Oxford: Oxford University Press.

Index

Printed in the United Kingdom by
Lightning Source UK Ltd., Milton Keynes
139342UK00008B/1/P